Orphan Black

Orphan Black
Performance, Gender, Biopolitics

Edited by
Andrea Goulet and Robert A. Rushing

intellect Bristol, UK / Chicago, USA

First published in the UK in 2018 by
Intellect, The Mill, Parnall Road, Fishponds, Bristol, BS16 3JG, UK

First published in the USA in 2018 by
Intellect, The University of Chicago Press, 1427 E. 60th Street,
Chicago, IL 60637, USA

A catalogue record for this book is available from the British Library.

Copy-editor: MPS Technologies
Cover designer: Aleksandra Szumlas
Indexer: Max Alvarez
Production manager: Naomi Curston
Typesetting: Contentra Technologies

Print ISBN: 978-1-78320-922-4
ePDF ISBN: 978-1-78320-924-8
ePUB ISBN: 978-1-78320-923-1

Contents

Acknowledgements vii

Introduction 1
Andrea Goulet and Robert A. Rushing

Part One: Performance/Technology/Gender 23

Gesture in *Orphan Black* 25
David F. Bell

Playing with TechnoDollies: The TV Actress and Other Technologies 39
Christopher Grobe

Animating Cloning: Special Effects and Mediated Bodies in *Orphan Black* and *Jurassic Park* 57
Simon Porzak

Watching While (Face) Blind: Clone Layering and Prosopagnosia 77
Sharrona Pearl

Part Two: Reproduction/Biopolitics/Community 93

Game of Clones: *Orphan Black*'s Family Romance 95
John C. Stout

Orphan Black and the Ideology of DNA 111
Hilary Neroni

Being Together: Immunity and Community in *Orphan Black* 127
Jessica Tanner

The Dancing Women: Decoding Biopolitical Fantasy 145
Robert A. Rushing

The Replicant's '*Réplique*': Motherhood and the Posthuman Family as Resistance in *Orphan Black* 167
Andrea Goulet

Afterword: Reflections on the Show, and Interviews with Cast, Crew and Creators 185
Lili Loofbourow

Appendix: *Orphan Black* Episodes 203

References 207

Notes on Contributors 217

Index 221

Acknowledgements

My thanks go first to my co-editor Rob Rushing, and to our shared diasporic UIUC community of colleagues and friends; I feel incredibly lucky to have begun my career in such an enriching intellectual environment. I am also grateful to the support I have received at the University of Pennsylvania, from individuals in my department as well as colleagues involved with the Mellon-funded Humanities + Urbanism + Design interdisciplinary faculty programme. Rob and I both appreciate the professionalism of the editorial staff at Intellect Books and would like also to thank the external readers of our manuscript for their expert advice. And finally, a note of gratitude to my family – Jed, Jonah and Maya – for their patience with someone who tries to embody the best bits (without too much of the craziness) of Alison, Sarah, etc. Clone Club, this volume is dedicated to you!

A.G.

I'd like to thank the very large community of family (especially Lilya and Sasha), friends, colleagues and students – *sestras*, all – who have joined me in watching and thinking about *Orphan Black*. I've twice had the good fortune to teach the show, once to graduate students in a seminar on film, television and biopolitics, and once to undergraduates as part of the University of Illinois Campus Honors Program. Both times I managed to make quite a few 'converts' to the show (Helena will do that to people), but as with all good students, I learned more than I taught. This is even more true of our fantastic contributors, whose delightful work I'd like to acknowledge here, as well. And, of course, I'd like to particularly thank my co-editor Andrea, who suggested this project in the first place, and who led the way to a panel at the American Comparative Literature Association's annual conference. This is our second collaboration in the last fifteen years – may it not be our last!

R.R.

Introduction

Andrea Goulet and Robert A. Rushing

Orphan Black begins with a twist. Anglo-Canadian punk-rocker and occasional con artist Sarah Manning is waiting on a train platform when she notices a well-dressed woman behaving suspiciously. She watches as the woman paces nervously, removes her high-heeled pumps, and places her purse against a pillar in the middle of the platform. When Sarah realizes that the woman is about to leap in front of an oncoming express train, it's too late to stop her. In the moment before the woman leaps, the two make eye contact for the first time, revealing something apparently impossible: this stranger has Sarah's own face. Over the next episodes, Sarah flees her former life as a petty criminal and her incompetent drug-dealing boyfriend by taking on the identity of her previously unknown twin, Beth Childs, a well-off policewoman with a hunky boyfriend named Paul.

This twist may be surprising, but it is hardly unprecedented for contemporary television viewers. Just two years before *Orphan Black* premiered, the series *Ringer* (2011–12) had featured Sarah Michelle Gellar as Bridget, a recovering drug addict and former stripper with a twin whose mysterious suicide allows Bridget to insert herself into her sister's wealthy and upstanding life. The shared premise of the two shows requires a certain degree of what we might call 'meta-acting', or acting about acting, since the actor (Maslany, Gellar) must play not only two separate characters, but also one character acting as the other. Of course, the premise also gratifies a fantasy about class mobility for young women: it is not an accident that both shows centre on the sister who is down and out (and single) moving into a life of comparative wealth, social status and romantic-sexual satisfaction. Even so, *Ringer* met with only middling reviews (although Gellar was generally praised for her portrayal of the twins) and was cancelled twice, while *Orphan Black* earned a steadily growing reputation and a good deal of popular press coverage, culminating in Tatiana Maslany's Emmy win for Best Actress at the end of the show's fourth season. The reasons for *Orphan Black*'s success and *Ringer*'s failure are no doubt legion, but it is worth noting that although *Orphan Black* begins with the same premise, it immediately multiplies it: yes, Sarah Manning discovers that she has a twin sister, Beth Childs, but she soon – and surprisingly – meets another 'twin', the German Katja Obinger; then the Canadian soccer mom Alison Hendrix; the Bay Area scientist Cosima Niehaus; the mad Ukrainian assassin Helena; the list goes on. If, on the one hand, the show exacerbates the 'twin sister' premise, it also deforms it. Unlike in *Ringer*, *Orphan Black*'s clone sisters are not merely somewhat different – they are radically different in appearance, personality, beliefs. *Orphan Black* weighed in early and often on the nature versus nurture debate, always taking the side of the environment's role in shaping the individual, emphasizing two ideas from Darwin's *The Origin of Species* that also appeared as episode titles in the first season: 'Variation Under Nature' and 'Endless Forms Most Beautiful' (see Neroni in this volume on *Orphan Black*'s opposition to the ideology of DNA as our 'true self').

As a result of this emphasis on expressive diversity, *Orphan Black* rapidly turned into a playground for virtuoso performances about performance. Tatiana Maslany's individual creations were such complete individuals that it was often difficult to remember that they were played by the same actress (see Pearl's and Bell's essays in this volume). Many of the clones are called on, at various times, to impersonate the others, requiring a kind of second-order virtuosity from Maslany, who must play a role on top of a role; these 'clone swaps' are successful, but not perfectly so, since some of the traits of the clone who is impersonating her sister should remain. It is, if you like, a deliberately slightly awkward performance. And this kind of performance can be varied, multiplied. Indeed, in the fourth episode of the first season ('Effects of External Conditions'), there is a sequence in which Helena poses as Sarah posing as Beth, and these 'clone swaps' quickly became a favourite with viewers.[1] Indeed, everyone on the show imitates, performs: Cosima briefly impersonates Dr Leekie in 2.3 (throughout this volume we will refer to episodes thusly, by season number and episode number; an appendix at the end of the volume lists all of the episodes and gives important information such as director, screenwriter and original air date); Felix pretends to be straight in 3.8; Mark poses as Rudy in 3.10; Donnie pretends to be Felix's gay partner in 4.4; Siobhan Sadler (better known to viewers as 'Mrs S' poses as a psychologist in 5.4 and so on. (The uninitiated shouldn't worry: these characters will be introduced in detail later.)

There is another layer of performance at work in the show, this time technical, and apparent in the scenes in which the clone sisters interact with each other. In the past, such doubling of actors relied largely on split screen technology, which required that the actor gives two performances that remained also physically separate. (Such scenes often have a forced artificiality to them.) But as two of this volume's contributors (Grobe and Porzak) discuss in some detail, Maslany has also mastered a different, third-order kind of performance, in which she acts with a body double and special technology that repeats camera movements identically take after take, allowing the footage to be put together later digitally so as to create a scene in which Maslany appears in multiple character roles, each interacting with the others in complex and highly realistic ways. The finale of Season 2, for example, famously concludes with a 'clone dance party' in which all four of the principal clone sisters as well as Felix (Sarah's foster brother) and Kira (Sarah's young daughter) meet in Felix's apartment and dance together. All the bodies and objects interact with each other, showcasing not only each clone's absolutely individual style of dancing but also Maslany's (and her body double, Kathryn Alexandre's) ability to replicate the same gestures and movements so that there is no sense of artificial distance or forced awkwardness.

The problem with virtuoso performance, of course, is that it can turn into a kind of party trick, a mere imitation game. *Orphan Black* grounds its obsession with performance, however, in a series of concerns that not only give the show a kind of gravitas, but that render those same virtuoso performances meaningful – even necessary and urgent. In particular, the show is clear that performance, as Grobe points out in his essay in this volume, has a particular value and urgency for women. While men may need to perform certain roles under extraordinary circumstances (see, for instance, Matthew Rhys's bravura multiple performances as a Russian spy in *The Americans* [2013–18]), performance is an everyday necessity for women. Sarah begins the show as a sometime con artist, of course, but it is certainly suburban housewife Alison who most powerfully foregrounds everyday appearances and role-playing. We learn that each of the female clones has a 'monitor', someone close to her who observes and takes notes on her behaviour – but Alison is surrounded by an entire community of people who monitor her (and each other). Her neighbour and friend Aynsley in particular plays the role of the classic 'nosy neighbour', who ultimately acts to enforce certain standards of behaviour and appearance. (Aynsley's death is one of the clearest wish-fulfilment sequences in the series.) Sarah, undetected, surveils Alison driving her children to a soccer match; the police surveil the Hendrix home; an audience watches Alison as she performs in a community theatre; she is surrounded by friends, neighbours and her pastor, who have all been monitoring her behaviour, during an intervention; she is watched by those same neighbours when she has sex with Aynsley's husband in her minivan; she is forcibly placed in rehab and monitored to guarantee that she behaves herself and she even spies on herself with a 'nanny cam' in an attempt to prove that her husband is her monitor.

If not quite to the same degree, Sarah is also subjected to unusual scrutiny in her performances – as Beth Childs, she is photographed surreptitiously by Beth's boyfriend Paul, who has suspicions about her true identity; she is constantly subjected to study by her police partner Art and she must perform as Beth in front of a police board reviewing Beth's apparently unjustified shooting of Maggie Chen. One is tempted to ascribe this kind of official surveillance and inspection to the fact that Sarah is impersonating someone else (a police officer, no less), but in fact Sarah faces similar scrutiny in her life as Sarah Manning, too. Unlike all the other clone sisters (who appear, at the beginning of the series, to have been rendered infertile by the cloning process), Sarah has her own biological child – and she is constantly monitored and judged for her inadequate performance as a mother. Before the discovery of her sister-clones, Sarah had left her daughter in the care of Sarah's own former foster mother, Siobhan Sadler, and it has been almost a year since Sarah and Kira were together. To acquire custody of her daughter, Sarah must meet a variety of benchmarks established by Mrs S, and prove that she is worthy as a mother.

These performances (as wife, as mother, as woman) – and the concomitant social approval or, more often, disapproval that accompany them – are of course familiar to many women as part of the tissue and texture of lived experience. The principal clone sisters function in many ways as opportunities to think about different ways to perform gender, and the different ways that different gender performances are received. Whereas Sarah's performances are generally successful (e.g. she gets Kira back and successfully impersonates Beth), the audience for *Orphan Black* is clearly drawn to sympathize with Alison Hendrix's numerous – and generally spectacular – failures as a performer, most notably her on-stage collapse, from a toxic combination of pills, alcohol and guilt, during the musical *Blood Ties*. Cosima, the show's science geek, is more resistant to traditional gender performances, preferring an 'alternative' lifestyle (dreadlocks, marijuana, bisexuality[2]) that, while no less of a performance, emphasizes autonomy and freedom from social conventions. Helena, on the other hand, represents a much more radical departure from traditional norms of behaviour. A trained killer who initially hunts down the other clones, she is prone to violence and eats constantly, in a grotesque and slovenly fashion. In one sequence from the show's fourth season, we see precisely how Helena's performance is received by spectators within the diegesis. A happy suburban family on vacation drives through the snow in the forest when one of the children excitedly asks her parents to stop the car because she's seen a deer. The deer, however, has just been slaughtered by the pregnant Helena, who is dressed in the fur of the other animals she's killed; she carries the bloody carcass past the horrified family while the mother frantically tells the children to get away from the window. The sequence is played for dark comic effect, but it is clear that Helena, as extreme as she is, represents a possible source of viewer identification, a source of possible pleasure precisely through gender non-conformity.

At first glance, *Orphan Black* represents yet another televisual 'typology' of women, most famously realized in the four principal characters of *Sex and the City* (1998–2004) (although going back at least to *Little Women* [1868–69]) and parodied by later shows like *Girls*. *Sex and the City* offered the viewer four 'types' ripe for viewer identification: a central, largely unmarked, self-aware character (Carrie), a nerdy professional (Miranda), an uptight, social conformist (Charlotte) and an assertive non-conformist who flouts conventional behaviour (Samantha). *Orphan Black* does in some way replicate those four characters with the comparatively normal, but highly self-aware and central Sarah, a nerdy scientist (Cosima), an uptight social conformist (Alison) and a non-conformist who flouts conventional behaviour (Helena). But again, *Orphan Black* grounds these potentially superficial personality types in a much deeper logic and set of concerns.[3] From its beginning, the show has relentlessly militated against any notion of biological destiny; if it adopts the fourfold typology of *Sex and the City*, it is not only to go beyond it in degree (there is a world of difference between Samantha's non-conformist behaviour, largely limited to her aggressive and assertive sexuality, and Helena's), but also to multiply the positions available to the viewer, from the (seemingly) vacuous blonde hairdresser Krystal to the icy calculations of Rachel, from the transgender clone Tony to the disabled Charlotte. Indeed, more than twenty clones had been revealed by the end of the show's fourth season – and in the series' final episode, the total number of *sestras* is given as 274.

Ultimately, this multiplication functions less like an astrological table ('what sign are you?'), and more and more like an alphabet. It is perhaps not an accident that the original three members of Clone Club are Alison, Beth and Cosima, an alphabetical beginning that suggests that clonehood is something like an *ars combinatoria*. The four principal clones of *Orphan Black*, then, would represent all women (and indeed, all life) not insofar as all viewers are expected to 'find themselves' somewhere within one or more of the characters, but in the way that even a very small alphabet (such as the ones and zeroes used by a computer) can generate all possible sentences – an infinite variety of them. 'To combine is to create', says Dr Leekie in 2.1. Indeed, the four principal clone sisters might be best understood as an elementary alphabet of four: adenine, cytosine, guanine and thymine, the four nucleobases that make up all DNA, and whose combinatorial variation produces all of the 'endless forms most beautiful' of life.

Orphan Black grounds its virtuosic performances of character and impersonation in a network of very real-world concerns about gender: how women perform femininity and motherhood for a set of real and implied observers, how the roles available for performance are constrained and how liberation from those roles may or may not be possible. But those reflections are also grounded in a set of concerns about science and technology and how they impact ordinary bodies (especially female bodies) and lives – even life itself. French historian Michel Foucault argued that the modern era comes into being when the state recognizes that medical and scientific knowledge about populations represent a form of power – that indeed the two are inextricable, forming a nexus that he called *le savoir-pouvoir*, or 'knowledge-power' (see Foucault 1978). Foucault went on to argue that a hallmark of the modern era is the application of power (particularly state power derived from medical and scientific knowledge) to life, health, vitality and reproduction – what he called *biopolitics*. Indeed, control over women's bodies, health and reproduction is absolutely central to *Orphan Black*. Everyone wants the clones – to destroy them as religious abominations (one sect of Proletheans), to use their reproductive capacity as a glorification of a cult leader (a different sect of Proletheans), as examples of transhumanist philosophy made flesh (Dr Leekie, the Neolutionists and the Dyad Institute), for military uses (Project Castor), for corporate exploitation of genetic engineering (Topside), for the commercialization of *in utero* genetic modification (BrightBorn) and more. Effectively, the show has found a narratively engaging way of describing what is a commonplace of women's lives – the constant attempts to control, monitor and regulate their bodies, particularly their reproductive capacity. Two men enter a diner (Episode 2.1), and one somewhat cryptically asks the chef, 'are your eggs domesticated?' This strange if apparently innocuous question about natural, free-range eggs conceals a very different concern; he is in fact a Prolethean religious extremist, and the question is not really directed at the owner of the diner, but rather at Sarah who sits, terrified, in a booth – and perhaps, by extension, to the viewer at home, or even the entire social field. Are your eggs domesticated?

Each season of *Orphan Black* has made use of titles that are quotations drawn from a specific text. The first season was devoted to Charles Darwin's 1859 *On the Origin of Species*. That famous treatise's full title, *On the Origin of Species by Means of Natural Selection, or the Preservation of Favoured Races in the Struggle of Life*, not only establishes the centrality of evolutionary biology to *Orphan Black*'s scientific premise, but also hints at the violent, creepy underside of artificial selection at work in the corporatized laboratories of Dyad and Neolution. Season 1's episode titles take us on a journey of discovery about the clones' multiplicity ('Variation Under Nature', 'Variations Under Domestication') through the disturbing tangle of body-mod networks ('Parts Developed in an Unusual Manner') and into a final celebration of our heroines' 'Endless Forms Most Beautiful'. For the titles of its second season, *Orphan Black* makes use of quotations from the seventeenth-century philosopher and statesman Francis Bacon. Taken from texts that range from the philosophical and scientific, like *The New Organon* (1620), to the utopian political thought of *New Atlantis* (1627), these titles jump quickly into the realm of irony: 'Governed by Sound Reason and True Religion' exposes the dangerous fanaticism of the Neolutionists; 'Governed as It Were by Chance' leaves the clones struggling against external, deterministic forces; and 'Ipsa scientia potestas est' (knowledge itself is power) teases both viewers and protagonists with glimpses of narrative understanding while leaving the true power to those murky multinational forces in the know. The turn to the political is intensified in Season 3, whose titles make use of snippets from Eisenhower's 1961 farewell address to the nation, delivered at the height of the Cold War. Here, we enter the sphere of the military-industrial complex (a term that Eisenhower is credited with coining), with the military control of the male Castor clones project intersecting with Topside plots for domination. Themes for this season are decidedly threatening, from 'Certain Agony of the Battlefield' to 'Community of Dreadful Fear and Hate' and 'Ruthless in Purpose, and Insidious in Method'. Season 4 moves us into a more hopeful area of potential resistance to capitalistic biopower, with its episode titles inspired by Donna Haraway's work on posthuman feminism in the 1991 book *Simians, Cyborgs, and Women: The Reinvention of Nature*. That volume includes Haraway's famous 1984 essay 'A cyborg manifesto', a posthumanist reflection on technological intrusions into the body, their dangers and their feminist potentials (see Grobe and Goulet in this volume for reflections on *Orphan Black* and Haraway's feminist cyborg). With titles like 'Transgressive Border Crossing', 'From Instinct to Rational Control', 'The Mitigation of Competition', this season of *Orphan Black* grants the members of Clone Club areas of agency/power/negotiation that emerge not in spite of their 'monstrous' mutations but because of them. Finally, Season 5 takes its titles from the 1914 poem 'Protest' by the theosophical poet Ella Wheeler Wilcox. With its call for active resistance against 'the lawlessness of wealth-protecting laws/That let the children and childbearers toil/To purchase ease for idle millionaires' (32), Wilcox's poem affirms the rights and responsibilities of women/female clones/cyborgs (i.e. all of us); citing Wilcox, the series' final episode calls on the clones – and viewers – 'To Right the Wrongs of Many'. In short, the show has been at pains to establish an intellectual and political ground beneath its techno-thriller plot lines, as well as its own lineage and genealogy (as Stout, among others in this volume, demonstrates). That ground, however, also focuses prominently on the possibilities for feminist action within the domain of the biopolitical (see, in this volume, Goulet and Rushing), and more generally suggests a certain degree of political resistance (see Tanner). Sarah frequently expresses a desire to return to her ordinary life, but the show as a whole makes it clear that such a life – 'ordinary' and untroubled, in which the body is not controlled and surveilled – is not a real option for Sarah, nor for anyone else.

Intersections

While we've organized the essays in this volume around a number of key terms, we do not mean to suggest that they are unrelated to each other. On the contrary, the essays here

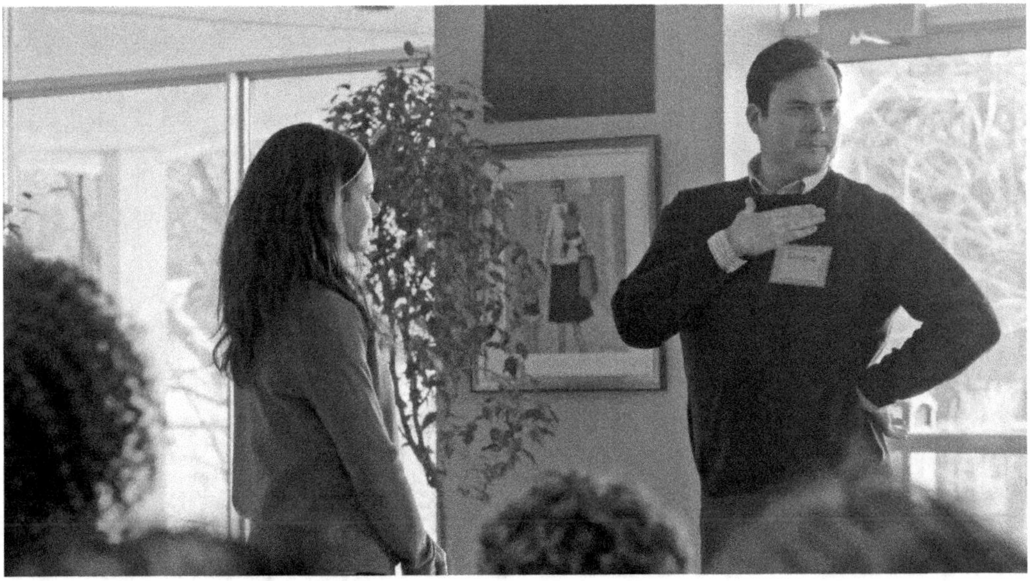

Figure 1: Donnie gets into character as Alison.

explicitly explore the overlaps between two or more of our key words. The interconnectedness of performance, gender, technology, reproduction and biopolitics can be glimpsed through a brief analysis of one scene in particular: the bravura comic performance in Episode 2.7 when Sarah is forced to pose as Alison in front of an audience of family members visiting their loved ones at a rehab centre. Here, Maslany must give a convincing performance not as Alison, but as Sarah pretending to be Alison. And in a typical *Orphan Black* performance twist, the organizer of this 'Family Day' event at the rehab centre abruptly asks 'Alison' and her husband to role-play *each other*. Maslany now performs Sarah performing Alison performing Donnie (Sarah briefly shines through: 'Oh, he's being Alison! And… [*slightly confused*] I'm being Alison being Donnie?'). But of course, the interchange is inevitably also about gender: Donnie gets into the role, adopting a caricature of Alison's typically prissy stance (see Figure 1 – note also the artwork on the wall behind the two, a mother taking care of two children while carrying a large shopping bag), and threatening to 'withhold affection' (Donnie speaks this last line knowingly to the audience, and getting a laugh at this recognizably 'wifely' manipulation through withholding sex).

Sarah-cum-Alison-cum-Donnie attempts to gain some leverage, saying that 'Donnie' wants Alison to leave rehab, but she's not really in character, and Donnie-cum-Alison outperforms her. 'No, Donnie, because I, Alison, need supervision', proclaims Donnie. At this point, however, Sarah/Alison/Donnie also gets into the spirit of things, and begins to perform masculinity. She relaxes her posture, spreads out, begins to take up space by swinging her arms and puts 'Alison' on the defensive by suggesting that Donnie – who really

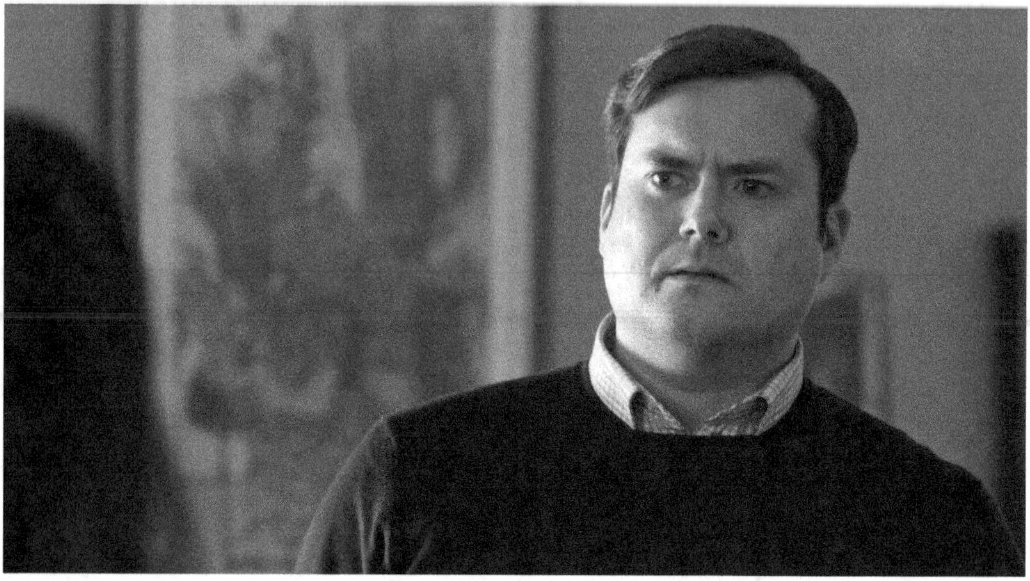

Figure 2: Donnie reacts to the disintegration of 'Alison'.

is Alison's monitor – spies on her. At this point, the flummoxed Donnie says, 'you're just using my voice for your words. [*Turning to the organizer*] Is she doing this right?' Moments later, Sarah's performance totally breaks down, as she refers to Alison as 'her' in a mess of unstable accents. Donnie looks alarmed as this total and uncanny breakdown of fixed, stable identity (see Figure 2). '*Now* who are you being?' he asks, unnerved.

There is, obviously, a kind of meta-performative delight in the sequence, since 'my voice' and 'your words' are both hopelessly confused ('my voice' is a phrase spoken by Donnie, but referring to Sarah pretending to be Alison pretending to be Donnie). We must also ask about the success and failure of performances: as the philosopher of language John Austin observed, certain sentences are true or false ('It's raining right now'), while other sentences are what he called 'performative utterances', like 'I now pronounce you man and wife'. These linguistic performances seem to either be successes or failures, rather than true or false (Austin's terms were *felicitous* and *infelicitous*). The felicity or infelicity of the role-play seems to be hopelessly unstable, however. It fails as marriage counselling for Donnie and Alison, but it is a success as well: Sarah uses the occasion once again to defend Alison against her husband – and she also uses the spectacle to distract the audience while the real Alison and Felix drag the unconscious Vic (Sarah's former boyfriend) down the stairs and out of sight. Gender is of course at stake in the scene and thrown into particularly sharp relief as Donnie and 'Alison' swap roles in a way that demonstrates how much their 'natural' genders are also performances that can be imitated and parodied. And finally, the biopolitical content is present in this scene: it is yet another occasion for the clones to bring out into the open

the fact that they are monitored for mental and physical health and stability (a monitoring that, in turn, demands another performance from them). That this biopolitical performance is itself gendered cannot really be doubted, either: with few exceptions, it is women being watched by men, and part of a more general social critique that is diffuse in *Orphan Black*, a critique of the constant scrutiny – social, legal, medical – given to women's bodies. And of course, the entire setting for this scene is a complex nexus of medical and legal power designed to help maintain and control health and vitality: rehab. This, in turn, is a space that represents what philosopher Giorgio Agamben (2005, making use of Carl Schmitt) calls the 'state of exception', a space in which normal laws, rights and freedoms do not apply. Indeed, Alison risks losing legal custody of her children if she does not complete the rehab programme – yet again an intersection of her gendered role as a mother and a biopolitical concern with life and health, tied to a demand for a successful performance.[4]

Almost all of *Orphan Black* functions like this, paying constant attention to the asymmetrical application of medical, corporate and legal power to the female body along with a concomitant demand that this body perform, both in order to comply with the demands of that biopower and to resist it. The intersection between gender and performance has, of course, been very widely recognized since the work of Judith Butler; and the intersection of gender and technology has similarly been much discussed in and since the work of figures like Donna Haraway. While this volume builds on those connections, it also aims to make use of *Orphan Black* to show the connections between gender and biopolitics, an area which has received comparatively little attention thus far. This is surprising, since the 'father' of biopolitics, Michel Foucault, was very much aware of the asymmetrical application of political, legal and medical power and scrutiny to certain groups (women, homosexuals, children). Much of the subsequent work, however, has tended to assume a universal (and thus, invisibly male) figure as the biopolitical subject, such as Agamben's *homo sacer* (see Rushing, in this volume). And the ways that biopolitics may be tied to performance are also largely ignored, even as they are manifest in our daily lives (the constant performance of health and vitality through fitness trackers, for example, or the social and professional performance of productivity). In short, this volume hopes to demonstrate that our key terms of gender, performance and biopolitics are ultimately not simply connected, but necessarily interlocking. If we can do that, however, it is only because *Orphan Black* has already demonstrated it so effectively.

One of the delights of the televisual format is that it creates a big, sprawling and complex world, one that allows the presentation and growth of highly idiosyncratic and complex individual characters. No single book can hope to exhaust everything that there is to say about a television series like *Orphan Black*, and this book does not want or pretend to do so. Among the evident topics elicited by the show that we do not address in depth in this volume is biomedical ethics. *Orphan Black* raises a number of questions very directly: can/should one patent a human being? Can one patent the *clone* of a human being? What are the limits of acceptable body modifications? (Most viewers seem to be fine with laser surgery to correct Cosima's near-sightedness, but less so with Olivier's fully functional and disturbingly phallic tail.) When should one intervene in genetic modification? What if such modification is prenatal? What if the modifications are made to the germ line, meaning that they will be passed on to future offspring? When we experiment on human subjects (by, say, cloning them), what are our moral and ethical responsibilities to those clones? These issues and many others have already been dealt with quite extensively in Gregory Pence's excellent *What We Talk about When We Talk about Clone Club*.

Another salient aspect of the show that our volume does not centrally address is *Orphan Black* as a specifically Canadian program. Indeed, *Orphan Black* is part of a burgeoning crop of Canadian science-fiction television series, including *ReGenesis* (2004–08) and *Between* (2015–16) (focused on biopolitics); *Dark Matter* (2015–17) (which also includes clones, albeit in a very different way) and *Killjoys* (a more conventional space opera); *Odyssey 5* (2002), *Continuum* (2012–15) and *Travelers* (2016–present) (which all tell the story of time travellers from the future sent back to the present day to prevent a catastrophe), and others. Any attempt to explore ways that Canadian science-fiction might differ from US science-fiction is somewhat impeded by the fact that *Orphan Black* seems at pains to create a diffuse and generic 'North American' setting for the show, likely to facilitate its export and distribution in other countries. While we do glimpse Canadian money and locations in Toronto, neither the country nor the city is ever named. In Episode 2.1, when Art is informed that federal agents have taken over a case, he asks, 'which Feds?' His partner replies, 'I didn't get an acronym', conveniently allowing the country to remain unspecified. At the same time, the show has also cultivated an international cosmopolitanism: we meet clones from Canada, England, the United States, Ukraine, Germany and Finland; viewers learn of the existence of other clones from Italy, France and Austria (and indeed comic book adaptations add clones from Poland, the Czech Republic, Denmark and the Netherlands); Rachel Duncan speaks fluent German (and even a bit of Korean); and Helena has taught everyone the Ukrainian word for sister: *sestra*. For the most part, *Orphan Black* as a global phenomenon (both in its distribution and in the globalization of its content) lies largely outside of the scope of this book. But it is our hope that the volume's areas of focus – performance/technology/gender and reproduction/biopolitics/community – will open critical avenues to these and other areas of inquiry. *Orphan Black*'s treatment of sexuality, for example, has been seen as both strongly progressive (Cosima's well-known affirmation that her 'sexuality is not the most interesting thing about [her]') but has also occasionally been critiqued as problematic (as was the case with Tony, the transgender clone); Felix, however, remains an under-examined element for the show: at times an independent and free-spirited gay artist, but at times perhaps sidelined as a variation on the 'gay male best friend'. One of the emergent themes over the course of the show's five seasons is Felix's attempt to establish a life, both professionally (his art career) and personally (through, for example, his discovery of his biological sister) – although both are continually entangled with his *sestras*.

The show's treatment of race is equally intriguing, and deserves further exploration. Some characters of colour such as (the largely underused) detective Art Bell are presented with no mention or attention to their race, as though it were not a factor – which is perhaps what one might expect from a show that champions nurture over nature. At the same time, however, there is a surprising amount of symbolic and racial blackness in *Orphan Black*. The origin of the Sarah and her *sestras* is certainly – from the show's title – black in some sense, facilitated in a profound way by blackness, including black characters such as Amelia, Sarah and Helena's birth mother, and Carlton, the agent responsible for smuggling clandestine children (clones and perhaps others) out of captivity. Certainly, it is no accident that reproduction throughout the show is subtly but pervasively linked to black bodies, both positively – Alison Hendrix's adopted children, Gemma and Oscar (Alison's last name, of course, links her also to Jimi Hendrix) – as well as negatively, where Alison's suburban community of Bailey Downs turns out to be quite racially fraught. Alison's mother, Connie Hendrix, initially dismisses the possibility that Alison and Cosima are even related, let alone clones. After seeing Cosima's dreadlocks, she announces with horror that 'that girl was a mulatto'. In that same episode, Alison offers a contrast to Connie's prejudice in a speech where she presents herself as a proponent of 'inclusion, not exclusion' – but her opponent in the school board race is the African-American Marci Coates, whom she defeats. Alison herself refers to Sarah's former boyfriend Vic as 'urban', a term that, when pressed, she admits means 'not white'; in 5.10, the series finale, she objects to the title of Helena's memoirs – she agrees they are all orphans, but 'we're not black!'. We might also note an implicit critique of racialized essentialism offered by the show in its fourth season, where once again it is a woman of colour, Kendra, whom we see as the latest surrogate mother utilized by Neolution (in this case, BrightBorn) to achieve its ends; the baby she gives birth to is not only white, but preternaturally so, with blind, milky blue eyes.

Organization and Contents

Our contributors' essays form a chain of resonances, but we have organized those resonances into two sections, recreating the quintessential *Orphan Black* notion of a split, the original 'twinning' that sparks the cloning process while also opening up to further multiplications, iterations – an originating (cell) division, in other words, that refuses to be reduced to a static binary. That logic is embodied visually in the show's opening title sequence (Kevin Chandoo, creative director), which features a series of twinned or mirror images – of Sarah's head, of DNA strands, of the bilaterally symmetrical orchid – that expand outward, giving us a hint of Sarah's discovery of one twin, then another and so on. Meanwhile, even the proper names of the cast and creators are 'split' by a vertical bar, as though to remind us of a central *clivage* in subject identity. Although we have divided the book's essays into two sections, then, they are connected by a number of overlaps and echoes. This is really more of a chain of terms – *performance/technology/gender/reproduction/biopolitics/community* – a chain that might suggest the complex forms of genealogy and filiation that are always at work in *Orphan Black*: the biological families, adoptive families and elective families that are featured in the show, as well as links to the television, literature, comic books and films that helped shape the series itself.

In the volume's first section, scholars in the fields of performance and technology studies explore the digital innovations and technical interactions between human and machine that allow the show to challenge conventional notions of performance and identity. In the book's second part, our contributors address family themes and *Orphan Black*'s own textual genealogy within the contexts of (post-)evolutionary science, reproductive technology and the politics of gender, while extending that inquiry on family to the broader question of community in a 'posthuman' world of biopolitics. Here, scholars mobilize philosophy, history of science and literary theory to analyse how *Orphan Black* depicts resistance to the many forms of power that attempt to capture, monitor and shape life.

Part One: Performance/Technology/Gender

Our volume begins with essays concerned with bodies – or, more precisely, with the question of embodiment: embodied performance. In the context of a twenty-first century transdisciplinary 'neuroscientific turn' (Littlefield and Johnson 2012), it is impossible to think about identity without acknowledging the dazzling complexity of interactions between brain, mind, sensation, expression and bodily gesture that go into the making (and perhaps the myth) of recognizable and autonomous personhood. If the ethics of epigenetics have given us new ways to consider determinism beyond Darwinian biology (Racine and Zimmerman 2012), clone fictions such as *Orphan Black* allow us to explore, through visualization, the multiply determined biofeedback loops between actor and spectator; this works intradiegetically, as we've seen in the theatrical scenes mentioned above (reactions in 'clone swaps'), and extradiegetically, through reported effects on viewers. The well-theorized notion of performativity allows us to say simultaneously that Tatiana Maslany is and isn't one person. And her performance(s) are interpreted, of course, through multiple screens, both technological and biological.

The opening essay, David F. Bell's, 'Gesture in *Orphan Black*', takes up the topic of bodily performance through a look at the ambivalent nature of gesture. Because it straddles the line between the conscious and the unconscious, gesture constitutes a significant entry point for analysing the notion of embodied cognition. The task of playing multiple roles imposed upon Tatiana Maslany in *Orphan Black* provides a fascinating test case for considering the nature of gesture not only in terms of viewer recognition, but also as a way to consider the stakes in an ongoing nature/nurture debate. The actor must give life to diverse versions of the clone characters, each of which results from the different social context in which any particular clone evolved. External elements like hairstyle, clothing and social context create credible manifestations of differences among the clones, but there are less obvious strategies in play when cinematic techniques of caching (hiding the use of split screen technology and multiple performances) and layering interact with bodily movement to create the illusion of 'organic' authenticity. In this chapter, Bell situates *Orphan Black* within a long genealogy of filmic manipulation, from Georges Méliès's illusions of dismemberment to Jean Renoir's counter-shot montages to *The Parent Trap*'s relatively clumsy doubling in 1961 of one actress playing twins. How, he asks, do technical strategies available to the show's producers today continue to reach (impossibly) towards perfect simulations of the embodied nature of social interaction? How does Maslany use gesture in both subtle and not-so-subtle ways to create the illusion that the clones she is playing are distinct in particular ways? And how is this process related to cognitive systems that respond to the gestural behaviour of the other? Ultimately Bell calls upon current neurocognitive theory to argue that the unconscious and social origins of gesture within individual beings make gesture an element that is not easily manipulated by actors; they cannot ultimately escape from their own gesture set, precisely because it is formulated during the earliest developmental stages and is thus unavoidably ingrained in their own bodies. But the visual technologies of television and cinema 'gesture', as it were, towards a fundamental – and fundamentally ambivalent – enactment of bodily gesture.

The next two essays in the volume continue to explore the interaction/interface of body and machine, with an emphasis on technologies through which *Orphan Black* heightens the theatrical and cinematic processes that blur the line between what is human and what is not. 'Playing with TechnoDollies: The TV Actress and Other Technologies' by Christopher Grobe, asks whether human bodies are 'vital organisms' or 'biological machines'. This has been the central question of acting theory from Diderot to Stanislavsky, but in an age of cybernetic living, the question hardly makes sense anymore. Bodies and machines are entangled, and thus, Grobe argues, our notions of 'the human' must change accordingly. Popular acting has always been a place where such changes are publicly negotiated. When actors decide to take on a new machine as a scene partner, this technology requires new techniques from the actor, and new *technês* of the human are made palpable. In other words, actors do not just test the technological affordances of the machine; they also strain at the limits of their own performed humanity. Grobe shows this happening in *Orphan Black*, with Maslany hailed for feats of characterization whose relation to 'authenticity' needs to be thought through the cutting-edge technology of the TechnoDolly, which repeats the same camera movement and focus, allowing editors to stitch multiple takes together late into a dynamic, Maslany-filled master. In response to this robotic scene partner, Maslany must regulate her own body, 'hitting her marks' with near-mechanical precision. Another actress (Kathryn Alexandre), meanwhile, is the ghost in the machine – doing the scene-work with Maslany, but then stepping aside while Maslany plays each part into thin air. Grobe calls upon Donna Haraway's famous essay 'A cyborg manifesto' to argue that *Orphan Black* not only thematises the story of human clones as combining militarism and capitalism, it also extends the 'technological construction' of character to all screen acting and off-screen performance; we are all machine-bodies.

The conventional distinction between natural ('organic') and artificial ('mechanical') performance is similarly discarded by Simon Porzak, in his 'Animating Cloning: The Body's Special Effects in *Orphan Black* and *Jurassic Park*'. Porzak's essay considers the 'special effects' in the show, noting that, compared to a film like Spielberg's *Jurassic Park*, *Orphan Black* appears relatively 'low-res'. *Orphan Black* almost exclusively employs the 'trick shot' or composite image, a technique dating back to the dawn of cinema in the nineteenth century, to visualize its futuristic science. Porzak argues that the viewer thus lacks any obvious nucleus in the image in which to localize the place of science-fiction within *Orphan Black*'s visual universe, since he or she cannot divide the fictional near-futurism from the realist present. However, *Jurassic Park* – in which scientists use cloning to re-animate a brave new world of female dinosaurs – also displays an interest in more archaic film technologies, notably in a sequence in which the cloning science at work is explained through an informational video that intermixes hand-drawn animation and live-action video. These works, then, both suggest that the most science-fictional special effects lie not in the future, but at the dawn of cinema, as the interconnected technologies of splicing and cloning that fracture a putatively organic life into a series of provisional and mediated animations. Porzak traces this media archaeology of *Orphan Black* and *Jurassic Park* to suggest a new understanding of the audiovisual science-fiction aesthetic while also calling attention to their gendered interest in the female actress as a medium for understanding the more general 'animatedness' – a term combining the illusory performance of selfhood with visual technologies – of the human subject.

Finally, we turn to the embodiment of the viewer. In the case of one set of viewers, a specific type of embodiment – the cognitive disorder called prosopagnosia, or face blindness – further 'screens' the reception of *Orphan Black*'s multiply-differentiated clone bodies. In 'Watching While (Face) Blind: Clone Layering and Prosopagnosia', Sharrona Pearl explores questions raised by the experience of a community of *Orphan Black* viewers affected by prosopagnosia. For spectators who are unable to distinguish individual facial features, this particular television series would seem completely to resist comprehension; the faces of characters played by Maslany are not just similar but identical – and distinguishing each clone is fundamental to the story. And yet, a surprising number of viewers with prosopagnosia report that they find *Orphan Black* to be one of the easiest shows on television to follow. In her essay, Pearl argues that *Orphan Black* works for prosopagnosiacs because the embodied presentation of each clone is explicitly differentiated in highly specific ways – through hair and makeup, accent and gesture, and technical tricks – so that recognizable identity becomes, in the terms used by Deleuze and Guattari, a question less of 'faciality' than of the 'abstract machine'. Through a close reading of modes of on-screen differentiation at the level of gesture and embodiment, alongside the written responses of people with prosopagnosia, Pearl shows why *Orphan Black* is so popular among the face blind and what that might mean for the rest of us. She focuses in particular on the 'double layering' involved in clone swap sequences as a way to read the communication of distinct identity. We are meant in these scenes to know that none of the *'sestras'*, no matter how identical, can entirely disappear into another person. (And gender, here, matters, since the male Castor clones, especially those who have been raised in the same military milieu, are less clearly indexed as individuals.) Indeed, that is the point of the whole show: even clones are unique, individual and (ideally) unknowable. Pearl addresses the show's technical strategies (makeup artists, hairstylists, costumers, script writers, directors, sound mixers, a body double and camera operators and technicians) that help create the separate characters out of one actor and, in so doing, analyses the role of the face as a means of communicating and distinguishing character on screen. But more importantly, she directs us to reflect on what happens when that role is subverted. The chapter argues that shows like *Orphan Black* open up possibilities for a broader understanding not only of who can play which roles, but what the limitations of our raced, classed, gendered and (dis)abled bodies actually are – and are not.

Part Two: Reproduction/Biopolitics/Community

The essays in this volume's second section approach the concept of *filiation* from two perspectives: (1) extradiegetically, by tracing an intertextual genealogy or 'family tree' for *Orphan Black* that includes explicit references (such as *The Island of Doctor Moreau* [1896], discussed in Stout's chapter), suggestive allusions (see Rushing's chapter on Conan Doyle) and potential parallels (see Goulet's chapter on French sci-fi cloning fictions); and (2) intradiegetically, by exploring the show's thematization and problematization of family, inheritance/heredity and non-biological definitions of community. While exposing the ideologies of DNA (Neroni) and biopolitics (Rushing), the essays also explore the possibilities, opened up by *Orphan Black*, of affirmative responses to biopower through alternative communities of women, clones, 'rejects' and cyborgs (Tanner; Goulet).

In 'Game of Clones: *Orphan Black*'s Family Romance', John C. Stout explores the overdetermined recurrence of the theme of family in *Orphan Black:* the Leda clone *sestras*, the Castor clone brothers, the role of father figures such as Dr Leekie and Henrik Johansson, and, most troublingly, the many 'bad mothers' (Bonnie Johansson, Dr Coady, Susan Duncan…) who populate the show. To attempt to make sense of *Orphan Black*'s preoccupation with family, Stout appeals to Freud's notion of the family romance, introduced in his short 1908 essay. In the family romance, the child subject imagines him- or herself to be precisely an orphan, separated from illustrious parents and being raised by surrogates. Uncertainties about the nature and boundaries of the self, as well as anxieties concerning reproduction and origins, underlie the onset of the family romance. Such anxieties also arguably constitute an unstated source of what is driving the plot of *Orphan Black*, and help to explain the intensity of the audience's fascination with the show. In his essay, Stout examines several intertextual links between *Orphan Black* and four of its determining fathers and one mother: H. G. Wells's *The Island of Doctor Moreau*; Aldous Huxley's *Brave New World* (1932); Ira Levin's *The Boys from Brazil* (1976); Kate Wilhelm's *Where Late the Sweet Birds Sang* (1976); and *Orphan Black* co-creator John Fawcett's film *Ginger Snaps* (2000). Together, these (inter) texts form a complex web that depicts a dystopian vision of reproduction, descent and filiation; the 'family romance' of the orphan counterbalances these harsh models, promising a future union with the protagonist's true family and real parents.

Hilary Neroni's '*Orphan Black* and the Ideology of DNA' argues that the premise of *Orphan Black* lies in the idea that DNA is both an authentic marker of individuality as well as a code that can lead to new medical advances. Obviously, the gene and DNA actually exist – but our collective cultural belief in the *idea* of DNA now functions as an ideological structure. *Orphan Black*, she contends, provides an unexpected challenge to the fundamentally biopolitical ideology that surrounds DNA, the idea that the subject *is* his or her DNA, i.e. that there is no part of the subject that transcends that truth of the body. On the one hand, *Orphan Black* expresses our current fears, anxieties and desires surrounding the discovery of genes as the ultimate truth of who we are. On the other hand, the show fascinates the viewer on an aesthetic level with the multiple types of femininity and subjectivity that the different cloned women exhibit – despite their ostensibly identical DNA. Tatiana Maslany's performance, playing all the female clones, emphasizes each woman's singularity through her portrayals of different types of femininity and sexuality. As the story of their cloning and their biological predicament unfolds, the antagonism between their subjectivity (variable, free) and their status as a clone (static and identical) becomes more and more apparent. Crucially, when male clones are introduced, the diversity expressed by women disappears – the male clones are alarmingly similar to each other, since they were all raised in the same environment. Neroni argues that the women of *Orphan Black* experience directly a flaw in the reigning biopolitical idea of the body (a pure expression of DNA) as truth; beyond the category of 'body', there remains the idea of *the subject* as something more than simply body.

In 'Being Together: Immunity and Community in *Orphan Black*', Jessica Tanner begins with a scene from the *Orphan Black* episode 'The Antisocialism of Sex' (4.7). Beth Childs speaks to Sarah Manning from beyond the grave, imploring Sarah to resist following in her suicidal footsteps: 'bring us together, Sarah'. Sarah's alcohol-fuelled vision brings the show to a pivotal juncture, one that confronts two modes of narrative progression: will Sarah join Beth in opting for death, reproducing the past and the show's originary narrative, or will she diverge from that plotline, re-joining her '*sestras*' who live on? Sarah opts to dwell in the present, endeavouring to bring together the clones who are her genetic identicals. Tanner argues that *Orphan Black* offers a vision of female subjectivity that reconfigures the paradigms of individual (male) subjectivation and social integration that govern the late nineteenth-century European novel (echoed in the series by the uncanny persistence of the purportedly Victorian-era founder of Neolution, P. T. Westmorland). Drawing on the work of Donna Haraway, Roberto Esposito and other contemporary thinkers of immunity, community and biopolitics, Tanner's essay argues that *Orphan Black*'s 'Clone Club' stages an affirmative form of community that counters the negative immunitary logic of Neolution. Sarah's own biological and narrative immunity symbolically allows for a protagonist that is both diachronically and synchronically dispersed, calling into question the distinction between self and other that subtends traditional conceptions of identity as proprietary. In a narrative mirroring of Rachel's sinister plan to 'marry the best of both worlds' by joining bots and clones, merging genetic modification with evolutionary determinism, Tanner argues that Sarah's plot rejects consolidation in favour of a model of non-reductive incorporation, symbolically dispersing her fertile immunity into a radical form of collective autoimmunity or community.

Robert A. Rushing, in 'The Dancing Women: Decoding Biopolitical Fantasy', introduces the concept of 'biopolitical fantasy', that is, the idea that biopolitical projects are always animated and supported by a fantasy, generally that of the healthy, happy, safe and long-lived body. At the same time, television and film often draw on biopolitical themes that are themselves fantasies: melodramatic fantasies about lost lives and shattered families, and thrilling fantasies about biopolitical disasters from meteor impacts to zombie menaces barely averted. The classic thriller has generally coupled the image of the female body at risk, saved at the last moment by the energetic male hero. An early example, Conan Doyle's 'The Adventure of the Dancing Men' (1903), is an important intertext for *Orphan Black*, Rushing claims. 'The Dancing Men' combines Sherlock Holmes's capacity to defend England from colonial menaces with fantasies of male universality (the stick figures that make up the code in the stories are abstract human figures, and hence declared to be men) and of female victimhood. *Orphan Black*, however, 'decodes' this fantasy, both through its images (Helena repeats the uncanny stick figures of Conan Doyle's 'Dancing Men', but adds skirts to them, daring the police to decode them) and through literal acts of decoding that reveal the underlying fantasy of the thriller: the appropriation of the female body. This happens most forcefully when Cosima and Delphine discover that an inexplicable extra piece of the clones' DNA is in fact text written in amino acids: a patent, claiming ownership of the clones, their bodies and any 'derivative materials', such as their children. *Orphan Black* makes visible, then, how biopolitical fantasy is both gendered and concealed, but it also offers its own biopolitical fantasy when the clone Cosima, gravely ill, is temporarily returned to vitality and the clone sisters all dance together, suggesting both a complete 'decoding' of Conan Doyle (it is now the 'Adventure of the Dancing Women'), as well as a glimpse of how forms of solidarity may be essential to overcoming the forces that menace the sisters and their bodies.

The last of the essays in our second section, 'The Replicant's *Réplique*: Motherhood and the Posthuman Family as Resistance in *Orphan Black*', by Andrea Goulet, asks whether the thematization of biological reproduction in *Orphan Black* might constitute a challenge to biopolitical regimes through, paradoxically, a seemingly traditional emphasis on family. When the mother-figure 'Mrs S' assures Gracie Johanssen that 'fertile or not fertile, we're all human', she raises the hypothetical question of clone humanity while exposing a more pressing central preoccupation with reproduction in today's age of technological replication. Indeed, from the protagonist Sarah's near-oophorectomy to cult-leader Johanssen's breeding experiments to the US Army's human sterilization weapon in the form of male clones whose sexual contact with victims atrophies their ovaries, *Orphan Black* stages the scientific and ethical limit-cases of biological procreation and its suppression. It is Sarah Manning's motherhood, of course, that marks her as a science-flouting aberration among the sister-clones and that attracts the jealous fury of the steely, sterile Rachel Duncan – and in this character dyad of fertile heroine/barren bitch, we can see echoes of a number of intertexts. Goulet's essay frames *Orphan Black*'s reconceptualizations of the womb as textual matrix in relation to two of its French-language sci-fi precursors: Jean-Michel Truong's *Reproduction Interdite* (1989), in which a sadistic, sterile clone threatens the military health of her father's mass-cloning corporation; and Maurice Dantec's *Babylon Babies* (1999), which allegorizes the eclipse of the male subject by a polymorphous cybernetic carrier of cloned infants. The chapter draws upon recent work by thinkers like Irving Goh and Catherine Malabou to consider 'Clone Club' as a model for new forms of communal solidarity and political agency.

We were particularly delighted that Lili Loofbourow was able to contribute an afterword that brings together many of the threads explored throughout the volume, while also bringing in her experience interviewing the cast, crew and creators of the series. For many viewers and critics, it was Loofbourow's celebrated 2015 essay, 'The many faces of Tatiana Maslany' in the *New York Times Magazine*, that initially brought their attention to the show. Here, Loofbourow is able to bring those insights, initially offered after the show's second season, up to date with the end of the series. Loofbourow draws some surprising connections across the length of the show: she draws parallels between Helena and Rachel, for instance, that are both unexpected and convincing, and explores the particularly rich, multiple meanings surrounding Rachel's eye – its loss, its prosthetic replacement, and its second loss. Indeed, Loofbourow's original piece on Maslany's 'many faces' and her afterword in this volume both think through not only the show's rich thematics, but also broader questions about being controlled, tested and observed – perhaps *monitored* would be the best term – that once again bring together the fictional world of *Orphan Black*, the world of acting and celebrity inhabited by Tatiana Maslany, and the 'real world' of ordinary viewers: gender, performance, politics.

Notes

1 Identity-switching performances are not limited to the show, but also take place in its reception, with fans sometimes dressing and posing as all four of the principal clones.

2 Cosima is referred to in this volume variously as lesbian and bisexual, in part because, although the show's creators have described her as bisexual, Cosima calls herself a lesbian in Episode 3.7. In the case of a fictional world, it is not entirely clear whether the creator(s) or the created have priority.

3 One might alternately argue that rather than provide a typology of 'varieties of women', *Orphan Black* uses its four principal characters primarily to approach four different genres, two dramatic and two comic (see also Loofbourow's essay in this volume). Sarah is largely associated with the detective and action/adventure genres, while Cosima is principally situated within the traditional questions of the melodrama (unfulfilled 'true love' and a slow, lingering illness). Alison, of course, is firmly part of the satire of suburban middle class, especially post-*Desperate Housewives* (2004–12, another show that replicated the fourfold typology of *Sex and the City*), while Helena emerges from the black comedy of shows like *Dexter* (2006–13).

4 To be sure, the series does not address the intersection of performance, gender and biopolitics solely for the female body. In the fourth episode of the fourth season, Donnie

and Felix pretend to be a gay couple seeking fertility counseling at Lifespring; Donnie, as he does in 2.7, gets into the act right away with some limp-wristed gestures and exaggeratedly 'gay' acting. Felix – who showed off his straight chops in a much more nuanced way in 3.8 – corrects Donnie, pointing out that 5–10 per cent of the men in Bailey Downs that Donnie already knows are gay. In short, gay men don't always – or even usually – 'act gay'. 'Stop mincing', urges Felix, and indeed the episode's title, 'From Instinct to Rational Control', suggests that in fact a degree of self-conscious self-awareness is necessary for a performance of gender to be 'felicitous' (a word that comes from the Latin *felix*, a familiar name) or successful. It isn't what is always obvious, or even what 'feels right'.

Part One

Performance/Technology/Gender

Gesture in *Orphan Black*

David F. Bell

When I was 14, I had a crush on Hayley Mills. It was 1961, when the original version of *The Parent Trap* opened, and I went to see it repeatedly, charmed by the young English ingénue. There were no VCRs or DVD players or Netflix streaming then, and viewing meant paying the price of entry each time I saw the film. I was spellbound by a movie in which the same actress played the roles of twin sisters. Precisely what portion of the charm was generated by the features of the actress or by the magic of the caching technique, which allowed her to appear simultaneously as different characters within the same frame, I cannot say for certain now. My recollection is veiled by what I have subsequently learned about cinema and did not know at the time. Little did I realize then, for instance, that Bette Davis had already accomplished the feat of playing twins long before *The Parent Trap*. *A Stolen Life* (1946) featured her in the dual role of twin sisters in a film focused more on Davis's portrayal of the contrasting characters than on the cinematographic technique that so intrigued me in *The Parent Trap*. Davis was to reprise the double role motif in *Dead Ringer* (1964), with the emphasis once again on her bravura portrayal of twins with quite different psychological traits in a film that was viewed at the time as a vehicle for her acting gifts. But even before the Davis films, Danny Kaye had played twin roles in *Wonder Man* (1945), and Betty Hutton did the same in the 1944 film *Here Come the Waves*, and indeed the theme was already well established in the silent period, with Dorothy Phillips as 'twins' appearing together in trick shots in the 1919 *The Right to Happiness*. In short, there was a cinematic history of twin roles played by the same actor of which I was blissfully unaware.[1]

As for the younger me, I remember well that the cinematographic caching technique played a prominent role in the way I understood *The Parent Trap*. I did not have the slightest inkling then how it worked – I simply felt its effects. There was a long road to travel before I was to realize that far from being a technical discovery along an idealized evolutionary timeline in the history of film, caching was, quite to the contrary, at the very origin of cinema history. The radically darkened Parisian stage theatres of the late nineteenth century, where Georges Méliès learned his craft, were the sites of trick lighting effects, completely obscuring parts of the visual frame on stage to simulate, for example, guillotined heads separating preternaturally from their bodies. Pre-cinematic theatre already used caching as a key special effect, and the gestural violence of cutting the body into parts, made possible by the illusion, was a ready-made motif.[2]

The simplest caching effects achieved in early cinematography involved blocking out half of the frame by obstructing the camera lens and filming a shot with action in the exposed half field, then rewinding the film stock to the beginning of the shot and reversing the procedure, blocking out the opposite half of the lens and filming action in the other half field (hence the term 'caching' from the French verb *cacher*, to hide). The two half-frame shots combined into a full frame on the film stock create the illusion that the action in each half frame, temporally divergent in the filming process, is simultaneous.[3] There are important constraints on shots composed in this simplest of manners, and those constraints carry over into more complicated techniques. The background must remain completely stable. Movement in the scenery that disturbs the fit between split frame parts undermines the effect of simultaneity. Moreover, if shots using the same actor are at stake, a stand-in double is often positioned in the cached portion of the frame, because the naturalness of a supposedly concurrent social interaction between two people is otherwise difficult to simulate. Playing to a void while maintaining a sense of dialogue and interaction will falter at some point, but even the attempt to resolve this difficulty by using a stand-in double does not completely solve the problem. Something crucial about animate movement and gesture is at stake here. *Gesture is an intensely social interaction*: forced to imagine an absent other, or to interact with a substitute other in the form of a stand-in, an actor invariably misses the mark at some point. A perceptible artificiality and occasional timing faults haunt the frame – the ghost of a missing spontaneous social exchange.

Therein lies the rub, which was to become the centre of my reflections on gesture and movement in cinema: as much as I enjoyed watching sequences where caching made it possible to present twins played by the same actor on screen simultaneously, there were always moments when the illusion was broken. In *The Parent Trap*, twins Susan and Sharon, facing each other in synchronized shots, occasionally did not make eye contact in a way that matched what one might expect – the angle of the gaze was slightly off. Or a reaction gesture after a dialogue line was slightly ill timed and did not quite fit the flow of the conversation. Why? In 1961, I attributed those moments to errors, flawed synchronization, timing faults, technical glitches I thought could be corrected. Why had these slight fissures in the continuity of such scenes not been properly fused? It seemed easy enough to me at the time.

Fast forward to *Orphan Black*. And then imagine the star-struck teenager of 1961 morphing into an *Orphan Black* fanboy of 2017. Indeed, one of the ways of consuming a popular contemporary television series is to join the fan base. Loyal viewers exchange information and comments online, often imagining different narrative directions that the story lines could take and writing different scripts to explore them (see, for example, Jenkins 2006). For *Orphan Black* fans, Tatiana Maslany's evolving performance, embodying multiple clones, is a constant source of discussion and admiration. It far surpasses the extreme simplicity, in hindsight, of the twin characters in *The Parent Trap*. And the high-tech caching techniques employed by the producers to display the cloned roles simultaneously on screen also receive strong billing in fan discussions. Accompanied by these incessant social media exchanges focused on key features that characterize the series, I am plunged back into the fundamental elements that defined my early experience of *The Parent Trap* – but this time to the nth degree. Moreover, the *Orphan Black* producers, Graeme Manson and John Fawcett, have seemed hell-bent from the outset to highlight their own technical prowess for their fans, creating season-ending scenes that put as many of the clone figures into the same frame as possible: the dancing sequence at the end of Season 2, in which the four main clones (Sarah, Alison, Helena and Cosima) dance together in the same space (BBC America 2014b), or the dinner sequence at the end of Season 3, in which they are seated at the same table, have become iconic (BBC America 2015). Manson and Fawcett go further: they take an almost perverse pleasure in peeling back the surface and revealing features of the sophisticated manipulations that create the illusion of simultaneous presence of several clone characters on screen, posting behind-the-scenes videos on YouTube – the *Wizard of Oz* (1939) unmasked. There are no more secrets about how the caching is done. We are shown green screen shots onto which backgrounds will be layered in post-production, as well as the TechnoDolly robot mechanism with its mounted camera that allows complicated camera motion to be duplicated with great precision by repeating a computer-programmed sequence of movements as many times as necessary (including automatic focus and zooming). No less crucial are the simpler spatial marking techniques revealed in these clips, since they allow Tatiana Maslany to situate her body and gestures correctly in the complicated spatial arrangements and angles necessary to establish the continuity of the cached shots in which she plays two or more clone roles in the same frame (actually four in the two scenes mentioned above).

The spell created by my experience as a 14-year-old is broken, but simultaneously – and unexpectedly – intensified. The teenager knew nothing about the technical magic required to construct a video image in a frame; his present counterpart knows markedly more and revels in arcane information about the techniques employed. Could this be a parable about what it means to become a critical (suspicious?) reader of texts, or images? Pleasure in exposing the technological marvels that allow the intricate staging that is such a vital part of *Orphan Black* is an undeniable dimension of this most modern of viewing experiences. Appreciating the mechanics of the techniques, as they are exposed in various YouTube videos about the series, corresponds strikingly to the gratification produced by exploring the sophisticated literary mechanisms that induce readers to lose themselves in a narrative or poem. Importantly, fundamental lessons about human interaction lie at the heart of caching techniques. Understanding these is the fruit of a critical labour originally triggered by my fascination with *The Parent Trap*. When assembling the layers of images to create the synchronized shots that display multiple clones in the same frame in *Orphan Black*, the illusion of actual social interaction can never quite be fully realized. Re-creating the *embodied nature of social interaction*, the fact that interlocutors move in what one might describe as an elaborate gestural dance in the presence of one another, is an aesthetic goal that can be attained only imperfectly, despite the extraordinary richness of the means now available to the filmmakers.

One of the early YouTube videos posted by the series producers highlights the difficulties of staging social interaction in shots when Tatiana Maslany plays more than one clone role and when the clones exchange dialogue while engaging in complex movements (BBC America 2013a). At stake in the clip is one of the first scenes in Season 1 of the series when Maslany actually plays three clones at once in the same frame in the family room of Alison's house, where Sarah, Alison and Cosima discuss the dangers that confront them. Graeme Manson begins with a statement of principle: 'That scene where Tatiana is playing three characters within the same frame is not conducive to organic acting'. The composed shot features two stand-in doubles with whom Maslany must interact as she alternately plays the roles of Alison, Sarah and Cosima in a series of different takes. She must switch roles and do new shots with a stand-in double now placed in a position she herself has just occupied. Manson lists some of the elements that must be combined in such shots: 'Getting all the timing down in that motion control, changing her wardrobe, playing the whole scene again with all the timing, getting it right, playing the scene empty and then doing it again, switching all the wardrobe and everything and getting into the third character'. The caching and layering that compose the frames require not only timing, but the ability to play off the stand-ins to maintain a sense of continuity in dialogic exchanges. Maslany reinforces the description of the degree of acting difficulty. She must position herself with respect to an intricate series of 'marks', which indicate the precise angles of her body when she turns, where she should look, where she should move and stop. As she puts it, she must replicate the 'motions I'm supposed to do', presumably meaning both the movements of the characters through the frame and also the sets of gestures accompanying those movements: 'I'm talking to, like, a mark on the wall over there, and then I look over here and I'm talking to this mark [...] and I'm trying to, also, like, do the motions I'm supposed to do, be on my cues, like, it's very, very, very, very, very technical'. Manson reasserts the essential points in his own summary: 'The dialogue has to be matching, all the timing has to work. It's incredibly difficult and requires, like, extremely technical acting on her part'.

As we begin to unpack the comments by Graeme Manson and Tatiana Maslany accompanying this brief video, the first thing to note is the binary opposition proposed between 'organic acting' and 'technical acting' – a distinction excellently discussed by Grobe in this volume. 'Organic' indicates the tenor of a shot in which actors exchange dialogue and move in tandem in the presence of each other on the set and in the frame. It refers to something like the continuity that is possible when actors interact with one another, playing off each other as the shot unfolds. In the remarks by Manson and Maslany, the notion of the organic becomes an ideal towards which the filmmaker and actor strive – easier to approach when an actor plays a single role accompanied by another actor, or other actors, in a shot. 'Technical' acting, on the other hand, is required when filming shots without the benefit of the continuity of an unfolding dialogic relation in real time, and it incorporates the full panoply of sophisticated caching techniques: playing to the correct marks; turning in the right direction, at the proper angle, in order to acknowledge the presence of another character, who may be there in the form of a stand-in, or may not be there at all ('playing the scene empty', as Manson puts it); placing the stand-ins in positions that will later be occupied by Maslany herself as she embodies another clone; hitting the timing of the intricate movements, face offs and dialogue exchange exactly right – or at least right enough to pass the viewing test. To reach a level of seamlessness that fools the viewer gives pleasure to the crew, as Maslany candidly admits: 'But there's something cool about being able to sell that shot with the three people in it. You want to get it, like, perfectly right. It's a kind of totally unnatural way of working, but it's, like, finding the freedom in that is the challenge'. The key is to naturalize the tricks, and then, within the multiple constraints they impose on the filmmaker and actor, to find some freedom, as Maslany puts it.

Unspoken by Manson and Maslany in this clip is an extended debate about continuity and montage dating back to early cinema, just like the caching technique itself. Broadly speaking, assembling individual shots into sequences can, on the one hand, make the cuts between shots extremely visible. For example, if the filmmaker juxtaposes shots of an unfolding action with shots that suggest symbolic meanings, shots that do not belong at all to the setting or the developing diegetic action of a scene. What Sergei Eisenstein first called a 'montage of attractions' confronts the spectator with emotional shocks elicited by unexpected associated shots grafted into a sequence to suggest ideas beyond immediate actions. Or, alternately, the filmmaker seeks to minimize the shocks that occur while assembling the shots into sequences by limiting to the maximum extent the spectator's perception of shot-to-shot transitions. The goal would be an ideal of *continuity* in which the spectator would not notice that sequences are always composed of separate shots. Montage is located on a spectrum that spans the gamut from the highly visible (Eisenstein's emotional shocks, for example) to near invisibility (the continuity tricks that every film school teaches budding filmmakers). If montage can never completely disappear, it can be occulted if the filmmaker so chooses.[4]

The assumption at work in the intricate caching techniques used to construct the scenes in which multiple clones appear is that continuity, the very invisibility of the techniques, is the ultimate goal. But the artificial manufacturing of continuity in cached scenes comes at a cost and leaves traces. A quick detour into the French director Jean Renoir's film aesthetics, under André Bazin's guidance, highlights the problem. Bazin's analysis of Renoir's technique in *La Règle du jeu* (*The Rules of the Game*) (1939) goes to the heart of the complex notion of continuity. Bazin emphasizes the fundamental interplay of actors with each other and with their surroundings – in a supposed continuum – as extended shots unfold in *La Règle du jeu*:

> Technically this conception of the screen assumes what I shall call lateral depth of field and the almost total disappearance of montage. [...] [T]he *mise en scène* cannot limit itself to what is presented on screen. [...] The action is not bounded by the screen, but merely passes through it. And a person who enters the camera's field of vision is coming from other areas of the action, and not from some limbo, some imaginary 'backstage'. Likewise, the camera should be able to spin suddenly without picking up any holes or dead spots in the action.
>
> (Bazin 1992: 89, original emphasis)

Continuity in Renoir means favouring long takes and reframing (panning and/or rotating the camera) when switching points of view, rather than, say, shot-counter-shot montage.[5] Bazin emphasizes the problems created by montage:

> as cleverly as these separate bits [shots] are stitched together, they cannot fool the attentive spectator. The little moment of hesitation at the beginning of a first line, the little something in the fixed nature of the camera and above all in the framing, where nothing is left to chance – everything betrays the existence of a preconceived 'shot'.
>
> (1992: 89)

I am brought back to my early impressions of what I perceived as errors while viewing *The Parent Trap*, impressions now reinforced by the hyper technicality of *Orphan Black*: caching techniques, no matter how sophisticated, invariably exhibit their montage to the detriment of a continuum of social space that might allow for genuine gesture and interactive movement. Gesture is eminently social, and no matter what a director or producer does to emulate social interaction by assembling takes in cached shots, the outcome will fall short of what Graeme Manson calls the organic. If Jean Renoir preferred long shots and reframing over montage, it was ultimately because he had in mind a particular way of allowing his actors to work with each other within those shots:

> Whereas most directors try to convince the viewer immediately of the objective and psychological reality of the action and subordinate both acting and directing to this end, Renoir seems to lose sight of the audience from time to time. His players do not face the camera but each other, as if acting for their personal pleasure. One senses that they become their own private audience, enjoying little inside jokes among themselves.
>
> (Bazin 1992: 77)

Renoir's actors, in occasional brilliantly filmed moments, seem to forget the staged nature of the scenes they enact, slipping into what one might describe as a commedia dell'arte improvisational mode characterized by the spontaneity of their social interaction, imprinting on their gestures a distinct feel of verisimilitude.[6] Tatiana Maslany can never let herself go in this way, as she concentrates on remembering gestures enacted by her stand-in, on listening to previously recorded dialogue in an earpiece, on hitting her marks and 'do[ing] the motions' that match what has been filmed in previous takes.

The revelations in the brief clip featuring Manson and Maslany can thus be situated within a broader and older history of cinema during which the notion of montage has been debated and developed – at the very origins of cinema, just as the caching itself belongs to cinema's origins.[7] In another YouTube clip that elaborates on the techniques described in the clip just analysed, Maslany again tries to explain the layering of shots that produces the illusion of simultaneous presence of the clones in the frame, this time in terms that harken back to Méliès's theatrical experiments (BBC America 2013b). She describes the various takes that

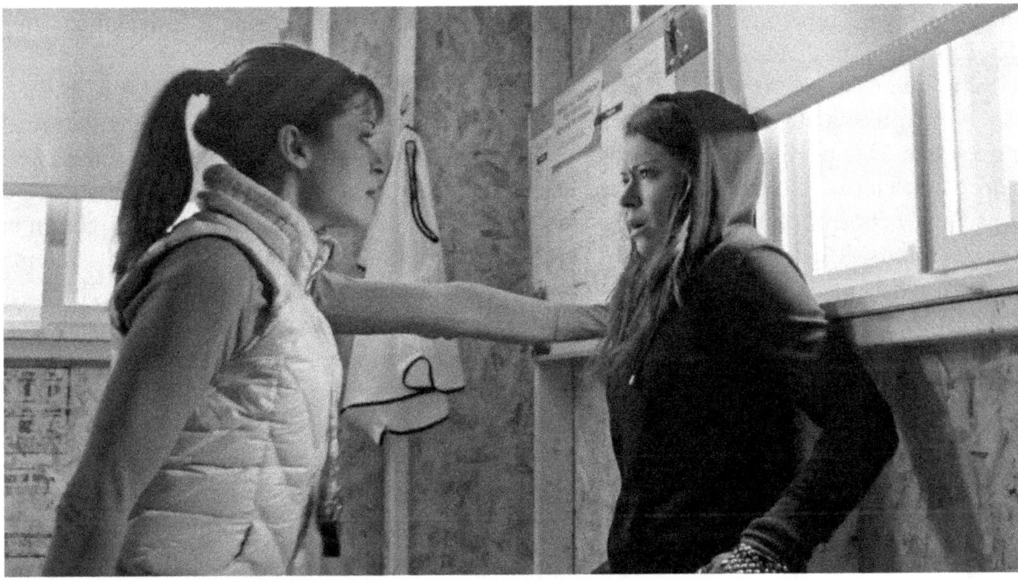

Figure 3: Digitally blended arm as clone meets clone.

compose the scene in Episode 2 of Season 1 when Sarah meets Alison for the first time in the shed next to the field where Alison's children are playing soccer. During an argument inside the shed, Alison tells Sarah to leave her alone and angrily pins Sarah against a wall (see Figure 3). Maslany comments on that moment, describing how the producers layered on parts of her stand-in's body over parts of her own body: 'they digitally cut her arm off here so that she's actually touching me but then my arm is from here up … Does that make any sense?' Yes, of course it does: Méliès knew perfectly well how to cut bodies into pieces on stage with convincing lighting tricks even before he began his career in cinema.

But why is gesture such a key element in 'selling the shot' as Maslany puts it? What is the social nature of gesture to which we have been referring? We tend uncritically to think of gesture as a type of bodily movement that accompanies the principal communication medium, language. Although it may even occasionally replace language in certain situations, it is nonetheless a supplement, we think, adding something to a linguistic exchange. What if we posit instead that gesture is not simply a supplement? Gesture is more essential to social interaction than common sense suggests, because it is first and foremost a manifestation of the mimetic nature of all animate social beings. The newborn is animate precisely because it moves from the moment of birth, even, of course, before birth, but at birth it immediately begins reacting to the movements of its caregiver by imitating them – well before it reaches a level of consciousness we later tend to imagine in Cartesian terms. Newborns do not 'think' and then move; they move in tandem with their caregiver, in unconscious imitation of the first person with whom they interact. The 'I reach' towards something can be explored analytically and attributed (partially erroneously) to conscious thought only much later (Sheets-Johnstone 2011, especially Chapter 3).

Subsequent stages in the elaboration of consciousness tend to reinforce a Cartesian mode of thinking about movement: we begin to believe that conscious thought 'controls' movement. This perspective obscures the immense repertory of movements accumulated before consciousness was fully formed and continuing to accrue until death stills us. We consciously control only a fraction of our gestures: the source of the remaining fraction is not conscious, if by conscious we mean decisions to activate our biomechanical systems to accomplish goals pre-programmed in our consciousness before they occur. Think of the gesture of reaching out to pick up a suitcase: if we had to calculate consciously the approximate weight of the object to be grasped every time we initiated such a movement, and thus the proper force, counterbalancing and angles necessary, the cognitive overhead would be enormous. But we already 'know' approximately what the weight is going to be and prepare the counterbalancing muscles and joint angles without conscious intervention, so that we can successfully accomplish the action. Hence those comical moments when we are wrong: the suitcase is empty and we apply too much force and stumble backward, or, alternately, it weighs considerably more than the roughly fifteen kilos we expected, our body's grasping gesture is completely stymied, and once again we stumble. Gestural movement contains unconscious 'muscle memory' accumulated over time, and this experience, formed through multiple repetitions of daily movement sequences, blends indistinguishably with whatever programme of grasping or moving we consciously initiate.[8]

A paradox thus lives at the heart of gesture as a result of the conscious/unconscious nature of the process of accumulating a repertory of gestures over time in social exchanges. On the one hand, gesture is highly mimetic, developed by imitating the movements of others and by repeating them until they become something like a 'second nature'. And yet the sum of our gestures is a decidedly personal 'signature' of our identity, a singularity recognizable by others in ways analogous to the graphologist's recognition of the particular handwriting of a given individual. Think for a moment of an instance when we might perceive someone walking ahead of us, at a distance and with a particular gait. We are able to recognize an acquaintance by the gait, even before we see the face, or, alternately, we mistakenly think the person is someone we know, because the nature of the gait seems akin to that of a friend. The way we hold ourselves, the way we move, the way we speak (our accent, our speech rhythms, our speech tone and so forth) are learned by imitating, but, simultaneously, the set of movements and gestures acquired through time is specific to us.

The mimetic dimension of gestures is what makes them difficult to deliver convincingly in cinematic situations such as the ones that occur in the multiple clone scenes of *Orphan Black*. Gestures are responses to the movements of our interlocutors; they are ways of connecting to and affecting others that go beyond language. Simply reciting dialogue in a persuasive and timely manner addresses only a part of the demands of social exchange in a scene when the actor is engaged with a fellow actor. The engagement is also something akin to a dance, during which the movements of one partner solicit responding movements from the other (Manning 2013, especially Chapter 1 on Argentinian tango). The failure to engage the look of the other, which so disturbed me at important moments in *The Parent Trap* when I was a teenager, is an artefact of the absence of mimetic interaction, generated by the absence of an interlocutor or by the presence of a stand-in who is not really the person addressed. The scene when Sarah meets Alison for the first time in the shed at the soccer field contains one or two almost imperceptible moments when Sarah's gaze misses Alison's. The casual viewer might not detect those moments, but the fanboy that I have become invariably stumbles upon them.

The notion of gesture at stake here is broad: it encompasses movements of the whole body, instead of simply the hand, head or facial expressions to which we too easily confine it. We sometimes recognize the holistic concept of gesture by referring to 'body language'. Tatiana Maslany leans heavily on this notion in order to distinguish among the clones she enacts. Broadly speaking, the more the gestures and movements are different and identifiable, the more they are recognizable, but simultaneously the more they tend towards exaggeration. The Sarah Manning character and the Alison Hendrix character are carefully contrasted through their respective gaits and bodily comportment from the very first time they meet in the sequence at the soccer field in Episode 2 of Season 1, discussed above. Sarah has followed Alison to soccer practice and gets out of her car to go towards the service shed at the edge of the practice field. With the help of slow motion in the first frames of the sequence, we notice that Sarah's gait is marked by a certain slouching and a rolling fluidity. At the end of the sequence that narrates their confrontation in the service shed, Alison returns to the soccer field, and her gait is substantially more upright and tight, with clipped steps. A brief final shot in the sequence comes back to Sarah, allowing us to compare the two. The pronounced tautness, nervousness and abruptness of the gait Maslany creates to portray Alison is even clearer in Episode 1 of Season 2 in a mall parking lot, when Vic sees Alison and mistakes her for Sarah, missing the gestural clues that might have disabused him.

The attempt to establish distinctions between the movements of the two characters tends, in fact, towards caricature. And caricature reaches a meta-level very quickly within the story itself, when Alison is called upon to replace Sarah to visit Sarah's daughter at Mrs S's house. Sarah had promised to come to the house in order to see Kira for the first time since her return to the city, but because she is unexpectedly forced to go meet Helena, she sends Alison in her place, and Alison 'becomes' Sarah for the first time. The recurring sequences in which one of the four principal clones replaces another become an important narrative motif as the series progresses. Felix coaches Alison in the taxi as they pull up to Mrs S's house, and when she exits the cab and starts walking towards the house, imitating Sarah's manner of holding herself and her gait, this elicits a compliment from Felix, but also a smile and a laugh, because the imitation exaggerates Sarah's movements. There is a detectable dimension of caricature in Alison's attempt to reproduce the bearing of her clone.

The feline manner that characterizes the Helena character is another example of a gesture set characterized by exaggerating certain movement traits. Because gesture is individually so deeply embodied, perhaps because it is a foundational source of cognition *tout court*, to create a subset of different gestures and movements to portray a different character ultimately requires exaggeration. Without strongly marked traits, actors inevitably revert to their own gesture set. This artificiality is visible enough to border regularly on caricature, and this very feature is incorporated into the *Orphan Black* narrative through both Alison and Helena, in particular, simply because it cannot be eliminated and therefore must be internalized into the narrative if it is not to risk the integrity of the whole. This should not be construed as a critique of Maslany's performance embodying the various clones, but rather as a way to understand the constraints under which she must work. Representing different clones is a paradigmatic case demonstrating the ultimate impossibility of making one's own body conform convincingly to a set of movements that are not one's own. Leave it to the child, Kira, not yet fully socialized, to recognize whose movements are genuine and whose are not. Kira's remark towards the end of the impersonation sequence in Mrs S's house, 'you're not my mother', goes to the heart of the matter. She is able to cut through the artificiality of the gestural repertory that Alison has attempted to construct in order to play her role as the absent Sarah. It is instructive to recall here that Heinrich von Kleist's 1811 essay, 'On the marionette theatre', ends with the ignominious defeat of an expert swordsman at the hands of a bear, who, like the child, is able instantly to distinguish movements that are feints from movements that are truly threats (Kleist 1972). The child and the animal are idealized figures of gestural movement untainted by accumulated experience that separates the body from the mind, obscuring the embodied dimension of cognition, which is constructed, at a very fundamental level, by movement and gesture.

A speculative conclusion concerning cinema and gesture beckons. Giorgio Agamben once made a deeply enigmatic remark about cinema: '*The element of cinema is gesture and not image*' (Agamben 2000: 54, original emphasis). Situated between acting as intervening in the world to produce a result and acting as embodying a character in a theatrical drama, cinema explores, precisely, mediality as pure mediality, gesture as what Agamben calls a means without ends, the exhibition of what cannot become a concept in language or an act in the world. Agamben calls this the *gag* in the two senses of the term: what prevents one from speaking and, also, the comic moment that crystallizes into gesture what cannot be reduced to language. The mask of the commedia dell'arte, he says, is neither a character nor a cluster of psychological traits:

> Harlequin and the Doctor are not characters in the same way in which Hamlet and Oedipus are: the masks are not *characters*, but rather *gestures* figured as a type, constellations of gestures. In this situation, the destruction of the role's identity goes hand in hand with the destruction of the actor's identity.
>
> (2000: 78, original emphasis)

In this cinematic space that Jean Renoir knew very well, gesture reigns: not the 'psychology' of the character or even the 'identity' of the actor, but rather the *enactment of gesture*. Small wonder, then, that the question of gesture is at the heart of *Orphan Black*, both in crucial moments of shot continuity, when cached shots exhibit the complexities of creating the gestural nature of social exchanges, and in the enactment of the clones by the principal actor in the series, heavily dependent on created gesture sets. A series about clones foregrounds Agamben's point wonderfully. The clones, clusters of gestures, break away from the actor by multiplying, thus destabilizing her as she grapples with the intricacies of playing them in the same frame. Is the essence of cinema truly gesture and not image? We need not see this as a binary choice. At the very least, however, we should take the lessons of *Orphan Black* to heart and recognize the key role of gesture in constructing the representation of interactions among characters within the frame.

Notes

1 My thanks to Robert Rushing for pointing out to me the Kaye, Hutton and Phillips films.

2 For an overview of the development of caching techniques in film over time, see, among many others, Rickitt (2000).

3 Given the array of sophisticated technologies available in contemporary video editing, a description of early caching technique that actually used film stock and simple lens blocking feels inevitably quaint.

4 Introductions to film theory inevitably elaborate on these contrasting approaches to montage (see Bordwell and Thompson 2012). Moreover, Eisenstein's montage of attractions is only one way of disrupting continuity. The filmmaker can also do this, for example, by violating certain rules of acceptable camera angle changes when cutting to the next shot or by abruptly and unexpectedly changing the location of the action.

5 A familiar example of shot-counter-shot montage is a conversational scene between two actors, during which the camera shoots over the shoulder of one of the actors to capture the face of the second actor, and then shifts to a shoot over the shoulder of the second actor in order to capture the face of the first actor, in a manner that appears to reproduce the back-and-forth of the conversation. In nearly every case, of course, the shots assembled to recreate the conversation are not filmed continuously and require the actors to recite their dialogue in pieces, at different times in a series of short takes. Hence the inevitable moments when the continuity of the conversation falters, revealing that the dialogue is composed of separate takes, of 'preconceived shots', as Bazin puts it. Such scenes are notorious for the continuity errors they introduce into the flow of the filmed conversation.

6 See, in particular, the soirée scene at the end of the film, which leads to the death of Christine de la Cheyniest.

7 See, for example, Chapter 2 of Williams (1992), where the Lumière brothers' 'realist' techniques are contrasted with Méliès's more contrived montage.

8 Honoré de Balzac, an early analyst of gesture and movement in his *Théorie de la démarche* (*Theory of the Gait*) (1832), recounts a practical joke he played on his sister Laure by overloading a sewing box with coins and watching her awkward stumble provoked by an underestimation of its weight as she grasped it. Gags based on misestimating the weight of some object are a staple of comic films based on slapstick physical humour.

Playing with TechnoDollies: The TV Actress and Other Technologies[1]

Christopher Grobe

Orphan Black is a show about women and clones. It is also a show about acting and labour. These facts are connected and, in this essay, I explore and explain a few of those connections. As you read, keep this endpoint in mind: the actor, the woman, the labourer and the clone are all alike. They are cyborgs. In fact, *you* are a cyborg, too – but let's not get ahead of ourselves.

The Amount of Acting in *Orphan Black*

> [...] we are not concerned, for example, with the degree of 'reality' but with what we can call, for now, the *amount* of acting.
>
> (Kirby 1972: 3, original emphasis)

More than any other programme in the history of television, *Orphan Black* begs its audience to pay attention to its acting. This claim may sound preposterous, but there aren't that many contenders. After all, the popular screen industries tend not to call much attention to the work that actors do – at least, not the kind of work that requires acting technique. (We're more likely to hear about the weight actors lost or muscles gained, or skills acquired in boxing or ballet, than to hear about new skills in acting, new approaches to building and sustaining a character.) What Cynthia Baron and Sharon Marie Carnicke say of early Hollywood could be said of television, too:

> [...] publicity surrounding screen actors has often actively suppressed information about training, preferring instead the myth of the born performer whose natural talents and genuine feelings are first captured by the camera, then presented on screen.
>
> (2010: 18)

The one exception proves the rule: Method acting has always piqued public interest, but only as some kind of anti-technique. Drawing on actors' 'genuine feelings' and life experience (sometimes experience they sought out for a particular role), the Method helps actors *be themselves*, even in character, even on camera – or so the story goes. There is a kernel of truth to this: like many modern acting theories, Lee Strasberg's Method is a *via negativa*. It is premised on the notion that, before true acting can begin, habits must be kicked, tensions released, conventions banished, body and soul restored to their factory settings. The 'amount of acting' must first approach zero – then rise, if ever, as little as possible.[2]

By contrast, the 'amount of acting' in *Orphan Black* was always conspicuously high. Fans of the show frequently measured this 'amount' – often by totting up the roles that *Orphan Black* asked its star to play. Even the media soon picked up this tic. Each year, around the time the Emmy awards were handed out, some journalist or other would write roughly this sentence: Tatiana Maslany 'doesn't just give one of the best performances on television – she gives nine of them, and counting' (Hiatt 2015). However glibly they meant it, people were offering this headcount as proof of Maslany's skill, reasons why she should win. Maslany is 'in another league', the *Washington Post* opined in 2014, 'as she plays, at this point, *nine* identical yet completely different clones' (Yahr 2014, original emphasis). And when she 'finally' did win an Emmy in 2016, the *Los Angeles Times* was ready to explain: Maslany was 'credited for more roles … than the rest of the lead drama actress nominees combined' (Brown 2016). One fan of the show – a data analyst by day – even measured Maslany's acting right down to the minute. This fan, Hannah Recht (2016), estimates that in its first four seasons (or 1658 minutes) on the air, *Orphan Black* gave its audience 1599 minutes of 'Maslanys'. In this same mood of straight-faced whimsy, so many people put a number – any number at all – to Maslany's performance.

But no matter how fun this may have been, and no matter how gobsmacking the numbers that resulted, none of this bean-counting was quite what theatre theorist Michael Kirby had in mind. What Kirby wanted when he set out to gauge 'the *amount* of acting' in different kinds of performance was 'a scale that measures the amount or degree of representation, simulation, impersonation and so forth' in various performance traditions (1972: 9). 'Not all performance is acting', he observes: certain strains of performance art, for instance, aimed instead for 'not-acting', where performers 'tended to "be" nobody or nothing other than themselves. … They walked, ran, said words, sang, washed dishes, swept, … and so forth, but they did not feign or impersonate' (1972: 3). Realist theatre, by contrast, might require these same actions yet turn each one of them into acting. Within the fictive frame of the realist stage, each quirk of behaviour must earn its place – must become, if it is to remain, stage business. Kirby's 'acting', then, refers to *fictionalized* behaviour, the concrete work of sustaining a fiction – and his 'amount' refers to the level of consciousness, choreography or choice that's evident in this behaviour: in short, 'feigning, simulation, and so forth that is *done* by a performer' (1972: 4, original emphasis).

Perhaps the reason people talk about the *quantity* of Maslany's acting is that they want – but don't know how – to discuss the *amount*. The two, of course, aren't unrelated. Precisely *because* Maslany spends so much time on screen, precisely *because* she plays so many different characters, you can't help but notice her acting – can't help but see how full of conscious choreography it is. When the camera cuts from one 'Maslany' to the next – in back-to-back scenes or in shot/countershot dialogue – it's impossible to ignore the distance travelled and all the 'feigning, simulation, and so forth' that happened along the way. And when *Orphan Black* shows us two or more 'Maslanys' in a single shot, it creates a rare opportunity: the chance to compare two distinct performances produced by one actor under identical conditions. Scenes like these turn entertainment into a well-controlled experiment, the casual viewer into a scientific observer. Even audiences unaccustomed to thinking this way will start to ask: are these characters fully distinct? Does Maslany inhabit each one deeply? Are they pliant personalities or stiff masks? What does she do (or fail to do) to make it so? In other words, the sheer *quantity* of her acting will make viewers start to notice what Kirby calls the *amount*. They will move from asking 'how much?' to wondering 'how?' – from matters of quantity to questions of technique.

The Labour of Acting

Though these questions feel inevitable to me, the writers of *Orphan Black* left nothing up to chance. They clearly relished the show's reputation as a showcase for acting – or, in the apt words of T. Colin Tait (2014), as a 'performance text of attractions' (*step right up, ladies and gents, and see nine roles played by only one woman!*) – and so they always looked for ways to raise the bar. They were, for instance, never shy to warp the plot if, in the process, they could create another showcase for Maslany's acting. They introduced new characters for her to play, they forced familiar characters into unfamiliar situations, and they required each of her primary characters to impersonate at least one of the other clones.

This last actorly ordeal, known by fans of the show as a 'clone swap', was a frequent feature of the show from the start. In the very first episode, Sarah Manning, the first 'Maslany' we meet – and, helpfully enough, a con artist by trade – must rehearse, then perform one of Maslany's own roles: the buttoned-up detective, Beth Childs. (In one especially 'meta' scene, we see Maslany *playing* Sarah *rehearsing* her performance as Beth so she can *practice* with Beth's partner for a high-stakes *performance* in front of the police department's conduct review board.) By the end of Episode 2, Sarah has added another role to her repertoire: that of her German counterpart, Katja Obinger. And in episode after episode, season after season, these 'clone swaps' add up: Alison plays Sarah, Sarah plays Rachel, Rachel plays Krystal, everybody plays Alison and (in one especially baroque variation on the theme) Sarah impersonates Alison during a therapy session where she must role-play as Alison's husband, Donnie. Not only are these scenes fan favourites – they were also a central plank in the show's marketing campaign. In the run-up to *Orphan Black*'s final season, BBC America released a video of the 'Best Clone Swaps' so far, and the show's official Twitter account ran a poll: 'What clone swap do you want to see for the final trip?' (BBC America 2017b). The answer, by a wide margin, was 'Cosima as Krystal', which lends proof to Staci Stutsman's theory that these clone swaps 'call attention to the performative nature' of gender itself by asking characters to embody 'different types of femininity' (2016: 97). After all, asking Cosima (a brainy, bohemian lesbian) to play Krystal (a ditzy 'normal') arguably requires the biggest leap between any two femininities shown on *Orphan Black*.

The message of the 'clone swap' is clear: acting is work, but work related to the 'performance' that each of us gives every day – at work and at home, for others and for ourselves. Actors are people who turn this work into labour, and, in so doing, make the labour of life itself suddenly visible. So, the pleasure you take in watching these scenes – or in imagining as-yet-unrealized permutations on them – is a pleasure you can also take in life: you can feel your wits sharpen as you attend to more elements of identity than you ever noticed before. I, for one, was never so aware of Sarah Manning's crooked smile – or of the anxious lip-biting habit it reflects – as when I saw it crack the mask of Beth Childs or Katja Obinger, marking the end of another successful con. The show encourages this way of watching. When, for instance, Sarah's foster-brother Felix warns Alison that she's 'over-slouching' in the role of Sarah, we are meant to silently add one more item (*degree of slouch*) to our checklist of variables in human behaviour. If we do, we extend by one significant digit our measure of Maslany's acting, of Alison's act – even of Sarah's daily self-presentation – and our own. It may not seem like 'work' to Sarah, but for Alison (as for Maslany), Sarah's way of holding her body, of using her voice, of managing her emotions, of navigating the social world is, in fact, a kind of 'performance' and a complex form of labour.

This has led to a rare situation: broad awareness of acting, not as 'natural' behaviour caught on camera, but as the work of a well-managed body and mind. More column-inches have been devoted to Maslany's acting than to nearly any aspect of *Orphan Black*, and much of this coverage has focused, unusually, on her process of preparation and characterization. Most interviewers, though interested in this process, seem unaware of acting techniques beyond Strasberg's Method. Maslany is quick to enlighten them. Consider, for example, this exchange from an interview with *Vanity Fair*: the reporter asks, '[s]ince you're playing five different characters on the show and you can't fully immerse yourself in them, Method acting-wise, how do you engage with them?' Maslany's answer breaks the unspoken rule that screen actors must seem (and claim) to be natural wonders: 'I do acting classes all the time', she explains, especially the kind where you 'dive in and dive out' of character, like some kind of actorly 'gymnastics' (Locker 2015). This 'gymnastics' has little to do with 'genuine feelings … captured by the camera', and even less to do with Strasberg's Method, as it is popularly conceived – with artless actors turning inward to achieve 'private moments'.[3] Instead, it is laboured, embodied and relational.

You could see the consensus turn: acting was now, journalists agreed, a matter of complex physical regimens. Word spread quickly that Maslany got into character not through private meditation or psychological reflection, but by dancing to custom playlists whose songs she selected for their power to set one of these women onto her body, carrying a character's spirit down into her bones (Morris 2015). Soon enough, these dances sank past skin, nerve and bone to reach the murkier depths of Maslany's mind. 'I didn't sleep' much during the first two seasons, she confessed to Lili Loofbourow of the *New York Times Magazine*: 'I'd have to do shifts during the day where I'd be Cosima … and then Helena – or whatever, Cosima and then Sarah. So my body was physically shifting in my sleep and I could feel it' (Loofbourow 2015). How's that for an image of the actor's labouring body?

But Whose Labour – Whose Body?

If *Orphan Black* made us aware of all the work Maslany does, it also occluded the work of many others. As Zoë Shacklock (2016) has argued at length, Maslany depended first and foremost on her primary acting double, Kathryn Alexandre. Alexandre is 'incredible', Maslany explains in one behind-the-scenes video, 'because she's rarely seen on screen – her face is never seen on screen – and yet she does all the same work that I do' (BBC America 2014a). (*But don't tell labour lawyers or union reps.*) Both Maslany and the show's creators are careful to heap praise on Alexandre whenever they can. She is, they (and journalists) reflexively say, 'an actor in her own right' (BBC America 2014a; Loofbourow 2015). Perhaps in an attempt to make good on these words – or as a way of simply paying her more for her work – the show runners eventually cast Alexandre in a small role on the show and, starting in Season 5, began listing her as a 'guest star' for her doubling work in the post-show credits. Gracious gestures like these, however, hide an unsettling truth. First, Alexandre is an object: or, as she puts it herself, 'a helpful asset to Tatiana, so that she's not acting off a tennis ball [on a tripod] or a mark of tape on the wall' (BBC America 2014a). (These would be standard tools for aiming an actor's gaze and body language in the right direction.) Second, her body, in many cases, is nothing more than a stand-in or a template for Maslany's. In fact, in that same behind-the-scenes video, just after Alexandre calls herself a warm-blooded alternative to tape and tennis balls, the video cuts to a scene showing how this plays out on set. Maslany and Alexandre have just traded roles and are now working to create the second half of a shot where Maslany comes face to face with herself. Two men hunch over a monitor looking at a composite video in which the latest take is layered over footage from an earlier session, where the roles had been reversed. One man's finger is darting around the screen like a doctor explaining an X-ray: 'We got lip lineup, we got nose. [Tatiana's] eyes are dead-on where Kathryn's eyes are.' Yes, she's ready to be erased. The further you get from the set, the balder this treatment of Alexandre becomes. Describing a technically complex scene where one clone had to pour wine into a glass that another was holding, visual effects supervisor Geoff Scott explains, '[w]e took Katherine's [*sic*] arm at the sockets, and tracked that arm onto Tat's body.' Simple enough, but to complete the effect, 'we still had to sort of break her arm in a few places, shorten it, and reposition it' (Romanello 2015; see Figure 4). Cutting-edge editing techniques threaten to 'turn every … editor into a potential Dr Frankenstein', says James Naremore (1988: 25). That would make Maslany nothing but a monster, and Alexandre just a pile of parts from the morgue, but that's unfair to both of them: they are more than dead meat on an editor's slab.

After all, Maslany has won such acclaim because, even in the most violently edited of scenes, she manages to feel not monstrous, but human. As Loofbourow (2015) muses, 'without Maslany's nuanced performance', *Orphan Black* could have become something else: 'a show so bogged down in its technical ambition and so in love with the possibilities of its own technology that it seemed mechanical. Instead, the final product feels organic, natural, real.' If this is true, it is thanks to Alexandre, too. Maslany speaks of how her double 'brings a full life and a full performance … for me to play off of' – and Alexandre, for her

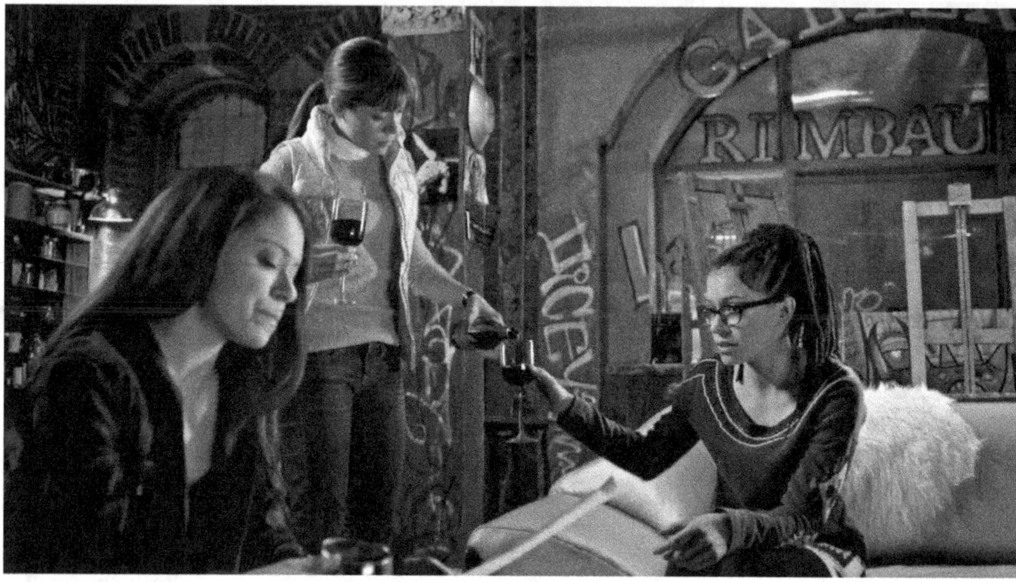

Figure 4: One clone pours wine for another.

part, 'always come[s] at it as an actor … throw[ing] her ideas on it', collaborating fully with Maslany, who then 'tweak[s it] from there' (BBC America 2014a). Even after Alexandre is erased, in other words, she's still there on the screen – and I mean more than her severed arms. If Maslany's performance feels psychologically deep, this is in part because someone is hiding in those depths. *Her* acting choices, *her* feelings, *her* ideas and *her* sheer reactive presence are sedimented deep within Maslany's performance. Alexandre is the ghost in the editing machine.

The Actor Is a Cyborg

> There is no fundamental, ontological separation […] of machine and organism, of technical and organic.
>
> (Haraway 1991: 178)

Two terms pervade our talk about acting today. They are so commonsensical, you actually read them in the last paragraph, and I doubt that you gave them a second thought: *technical* and *organic.*[4] Acting, it seems, can only ever be one of these things – or if it's both, these qualities are balanced *against* each other. This way of thinking has deep roots in western acting. In fact, according to Joseph Roach one question central to every western acting theory for centuries has been: Are human bodies 'vital organisms' or 'biological machines' (1993: 161)?

But I wonder: does it even make sense anymore to pose the question this way? Konstantin Stanislavsky – whose ideas inspired the Method, as well as other dominant theories of realist acting – could never seem to make up his own mind. He railed against 'mechanical acting' all the time, but wanted actors who were 'sensitive, responsive, mobile – like a well-oiled and regulated machine' (cited in Pitches 2006: 44). Likewise, Samuel Silas Curry, an American contemporary of Stanislavsky's, decried 'cold and mechanical' modes of performance, but taught performers to work on their own voices and bodies as an 'engineer does [on] his engine' (cited in McTeague 1993: 125, 132). The word for bad technical acting, everyone seemed to agree, was *mechanical* – but its opposite, though more organic, was also mechanical: the product of a well-tended 'engine', a 'well-oiled [...] machine'.

The choice between a 'vital organism' and a 'biological machine' now feels out of date in another way, too – a way that has more to do with changes in culture than in acting. We are all cyborgs now, 'creatures simultaneously animal and machine, who populate worlds ambiguously natural and crafted' (Haraway 1991: 149). Biology and vitality, animal and machine are so entangled by now, surely our notions of 'the human' must change – *have* changed – in fact, *are* changing all the time. These kinds of changes, when they happen, are always visible in the work of realist actors – those experts in the art of being human. When, for instance, people watched a nineteenth-century stage actor at the telegraph machine (Grobe 2016a) or a twentieth-century film actor on the telephone (Grobe 2016b), they learned something important. By studying those performances, so can we: we can learn not only how people welcomed new tools into their lives and their stories, but how humanity and technology were *intra-acting* at the time this performance occurred. I take this word from Karen Barad, a feminist philosopher of science who coined the term to do what *interaction* could not. When you say that two things *interact*, you presume a clear distinction existed between them from the start – but, as Barad points out, that's an odd thing to do. Just as the borders you see on a map were created (and are sustained) by all kinds of forces – mountains rise, rivers flow, nations fight, troops patrol, flags fly, citizens vote, people build walls (or *talk about* building walls) – so all kinds of difference, from normative gender to physical quanta, should, she argues, be understood as the result of an ongoing process, a 'differential becoming' (2003: 818). This outlook on the world, which Barad calls 'agential realism', might help us see what really happens when realist actors take machines for their scene partners. They stage a scene of intra-action where person and thing precipitate, as from a chemical solution. The world is shaken for a while, but then soon sifts out into new configurations of the human.

Orphan Black, in this sense, has something to teach us about humanity today, towards the start of the twenty-first century. In the show's fourth season, episode titles were drawn from the writings, not of Barad, but of a like-minded philosopher of science: Donna Haraway, a radical feminist and the author most famously of 'A cyborg manifesto'. When you hear the word 'cyborg', perhaps you see something closer to Paul Verhoeven's *RoboCop* than to

Orphan Black's clone/detective Beth Childs, but the borders between cyborgs and robots and clones have always been hard to draw. The word 'robot', coined by the Czech playwright Karel Čapek for his 1920 play *R.U.R. (Rossum's Universal Robots)*, means simply 'labourer' in Czech – and it originally referred not to clanking automatons, not to meat-and-metal cyborgs, but to a workforce engineered out of lab-grown flesh. As Domin, the manufacturer of robots in Čapek's play, explains to a visitor,

> [...] there are vats for livers, brains, etcetera. Then you'll see the bone factory, and after that I'll show you the spinning mill. [...] The spinning mill for nerves. The spinning mill for veins. The spinning mill where miles and miles of digestive tract are made at once.
> (Čapek 2004: 13)

Subsequent robot lore often does the same thing that *R.U.R.* does here, blurring the distinction between organic and synthetic creatures. Think, for instance, of HBO's *Westworld* (2016–present), whose title sequence shows a twenty-first-century version of Čapek's 'spinning mills' – they look a lot like 3D printers – and whose plot tells the story of what happens when a theme park's 'hosts' (a suitable synonym for 'robots' in a service job and in an age of affective labour) are given something more than flesh, more than artificial intelligence: an unconscious mind filled with Reveries™. What Čapek's 'robots' and *Westworld*'s 'hosts' have in common with cyborgs and clones – what always interested Haraway about 'cyborgs' in the first place – is their ability to kick dust over all the bright lines we might draw, or might mistakenly think were drawn by nature: the lines separating individuals from one another, subjects from objects, people from property and genomes from computational codes.

So, it should not be surprising that, even as *Orphan Black* tells the story of two sets of human clones – 'the illegitimate offspring of militarism [Castor clones] and patriarchal capitalism [Leda clones]' (Haraway 1991: 151) – it also tells a robot or cyborg story. You hear evidence of the show's robot roots every time someone talks of the clones becoming 'self-aware' – a word common in tales of artificially intelligent robots, but unusual in a story of human cloning. And these clones' cyborg souls are revealed when a mysterious bit of their genome turns out to be a stretch of binary code, a verbal message encoded in nucleotides and ASCII. These clones may be wetware – spun of sinew, nerve and bone – but they are robots and cyborgs nonetheless. As engineered creatures, they take part in that 'border war' declared by Haraway, joined by Barad and ongoing today (Haraway 1991: 150 and passim).

Characters Are Cyborgs, Too

While the clones of *Orphan Black*, considered as people, may be 'cyborgs' in this broadened sense, they are also, when considered as *characters*, 'cyborgs' in a narrower and more familiar sense. In a way, all film and TV characters are cyborgs: performed on-set by human actors, but engineered at many points between the set and the screen by many others, human and machine. Cameras and microphones capture the actor's performance, editors chop it apart and splice it together again, voices are dubbed, body doubles are swapped in and whatever remains is then altered by all sorts of software and hardware. In more extreme situations (think: characters created through 'motion capture' and digital animation, or with the help of 'face replacement' techniques) we see far more of these belated constructions than we do of the performance that 'originally' happened on set. Mark J. P. Wolf (2003) has given a name to this complex interaction of actors, technicians and apparatus: 'the technological construction of performance'. Each role requires 'an ensemble performance', Wolf argues, and each character is a composite creation (Wolf 2003: 55).

This idea doesn't sit well with some actors, who view performance through the lens of competitive individualism. Andy Serkis, an actor and a specialist in motion-captured performance, is a clear-cut case in point. Asked in 2014 about the state of the field, Serkis explained that animators, thank goodness, had been

> schooled [...] to honor the performances that are given by the actors on set. [...] It's a given that they absolutely copy [these performances] to the letter, to the point in effect [that] what they are doing is painting digital makeup onto actors' performances.
>
> (Woerner 2014)

The response from the visual effects community was instant and fierce. Cartoon Brew, a site for 'animation news', summed up the mood among digital animators with the following acid headline: 'Andy Serkis Does Everything, Animators Do Nothing, Says Andy Serkis' (Amidi 2014). Serkis's main motivation in saying this was perfectly clear to informed onlookers: he had long wanted greater recognition for his work – specifically, he wanted acting awards – but 'some people – namely, Oscar voters – have had difficulty accepting it as acting' (Robey 2014). In other words, the institution built to honour a 'best actor' (resolutely singular) didn't know what to do with an 'ensemble performance' like this, and so Serkis had to insist he was flying solo. He had to insist that the apparatus (and all the people who manned its cameras and computers) were transparent to his performance. If we accept Haraway's notion that the dream of our autonomy from other people, from nature and from machines is anti-feminist, then Serkis's fight for recognition is also, at heart, a gender panic. Perhaps that's why Serkis dismissed the claims of animators on his performance by trying to feminize their labour as the application of 'digital make-up'.[5]

In fact, women have always been central to the stories we tell about the 'technological construction' of character in film – and, just as Serkis evidently fears, these characters are often figured as the monstrous female victims of powerful male editors. Consider, for instance, the canonical tale of Lev Kuleshov's early experiments in film editing. In one experiment, Kuleshov created a film showing a woman 'sitting before her mirror, painting her eyelashes and brows, putting on lipstick and slippers' (Kuleshov 1974: 53). Despite this film's seamless appearance, Kuleshov had actually 'spliced' it together using 'the lips of one woman, the legs of another, the back of a third, and the eyes of a fourth' (Kuleshov 1974: 53). As Baron and Carnicke observe, '[f]ilm scholars tend to view this [experiment] as primary evidence that the montage, not the actor, performs' (2010: 34). In another experiment – this one the namesake of the 'Kuleshov effect' – Kuleshov stopped stitching women together from parts and started using them instead to stuff their male counterparts full of delicate emotion. Intercutting different secondary footage with the same primary close-up on a man's 'expressionlessly neutral face', Kuleshov created a performance that never happened on set: the audience 'marvelled at the sensitivity of the actor's [emotional] range', even as he tried to do nothing in particular (Kuleshov 1974: 8). The secondary footage Kuleshov used, his translator explains, showed 'a bowl of steaming soup, a woman in a coffin, and a child playing with a toy bear' – or, as Kuleshov recalled, 'a plate of soup, a girl, [and] a child's coffin' – or, as told on the dust jacket of the book from which I've quoted both versions, 'a bowl of soup, a child in a coffin, [and] a sunny landscape' (Kuleshov 1974: 8, 200, front flap). Landscape or person, 'girl' or adult, dead or alive, women were food for the soul of a sensitive man. They were meant to be sawn apart, ground down and rebuilt from parts as the film apparatus required.

But this 'technological construction' of character is usually so seamlessly done – or else its seams are so hidden by habit and convention – that casual viewers rarely know they are watching cyborgs. Most of the time, only industry insiders and film scholars give this matter any thought – like Lisa Bode who, in her study of 'posthumous performance' in film and on TV, envisions a 'continuum encompassing all the modes of technological mediation and augmentation of performance':

> It could range from more acting-centered performances, such as those captured by long takes, aesthetically centered on choices that actors make with their voices, bodies, and faces, to more overtly technologically bound performances [...] where details of the actors' recorded performances are reconstituted or repurposed within new mise-en-scène and interactional contexts.
>
> (Bode 2010: 69–70)

But *Orphan Black*, by cloning the same performer onscreen, makes even the casual viewer aware of – and curious about – the apparatus. (And once a fan does get curious, BBC America's YouTube feed will be ready to supply all the information they need.) The more deeply a viewer might probe into the apparatus of *Orphan Black*, exploring the techniques and technologies of onscreen cloning, the less Bode's 'continuum' will seem to explain. Like many scholars of film and TV, Bode presumes (if only rhetorically) a conflict between acting and the apparatus – a zero-sum trade-off where the amount of acting is always inversely proportionate to the complexity of the apparatus, but *Orphan Black* gives the lie to such assumptions. Where would it fit on Bode's continuum? Nowhere at all – unless we could twist it into a Mobius strip and run along its endless surface.

The TechnoDolly Is Us

> The machine is not an *it* to be animated, worshipped and dominated. The machine is us, our processes, an aspect of our embodiment.
>
> (Haraway 1991: 180, original emphasis)

After all, *Orphan Black* is one of the most 'acting-centred' shows on TV, but it is also one of the most 'technologically bound'. In fact, a good deal of the production team's technical ingenuity (and, no doubt, their postproduction budget) is spent creating 'long takes, aesthetically centered' on the acting of people playing two or more clones at once. In these long, generally low-key shots, the camera is almost always on the move – even if only slowly zooming in. But in some of the most intimate (and thus most engineered) of these shots, the camera moves freely, interacting (it seems) spontaneously with the actors – as responsive to them as a scene partner would be. Such shots are made possible by a cutting-edge 'motion control' system known as the TechnoDolly. Essentially, it's a robotic crane that can put a camera anywhere within a space about the size of a tennis court – but, more importantly, it can memorize complex camera routines and repeat them precisely, take after take – panning, tilting, zooming and refocusing the same way every time (Pacific Motion Control n.d.). This exact comparability across multiple takes allows editors to build up a shot in postproduction – layer by layer, as in an old-school split-screen or composite shot, except happening now in dynamic, three-dimensional space.

The TechnoDolly gives rise to a perfect paradox: its purpose is to enable some of the show's most intimate moments, but in the process, it requires actors to regiment their bodies with millimetre-perfect precision. You can hear the actors and show-runners trying to work through this paradox as they describe working together on these shots. They return again and again to those two master terms of contemporary acting: *technical* and *organic*. These shots exist, says *Orphan Black* co-creator John Fawcett, 'so that [multiple clones] can move organically within the frame and interact as though they [were] together' (BBC America 2013b). But this process, co-creator Graeme Manson observes, is 'not conducive to organic acting'; in fact, it 'requires, like, *extremely* technical acting on [Maslany's] part' – or, as Maslany herself puts it, 'I'm talking to, like, a mark on the wall over there', while also 'trying to […] do the motions I'm supposed to do, be on my cues. It's very, very, very, very, very technical' (BBC America 2013a). In short, to stay 'organic' while playing with TechnoDollies, you must become near-mechanical yourself. You must engage in 'very, *very* technical' acting.

Even as it technologizes the actors, though, the TechnoDolly also humanizes the camera. After all, it has the ability not just to run pre-programmed routines, but also to memorize shots that are live, human-operated and spontaneously adjusted. 'If you watch something that's just been created out of a computer, it can be very robotic-looking', John Fawcett explains, but 'when you're operating a camera manually, there are all these spontaneous moves and little imperfections in the camerawork that make it feel real' (McGeough 2015). In other words, it encodes a human operator's instincts, which, in turn, follow the scene first worked out by two actors, their director, and many other crew members.

The TechnoDolly, then, limits these actors not exactly by constraining their impulses, but by encoding them – inviting the actors to repeat them in take after take after take. Is this really so odd – or is it just an exaggerated description of all screen acting today? As TV scholar Richard Hewett puts it,

> While it might be thought that the possibility of additional takes provides the actor with more flexibility, enabling them to 'try out' different performance ideas [...], the majority of these takes in fact take the form of 'coverage' from different angles, in which the same moves and intonation need to be precisely replicated for the sake of editing continuity.
>
> (2015: 79)

Each new technology, says Lisa Bode, forces us 'to renegotiate [...] [the] contribution of acting vis-à-vis the contribution of technology and crew' (2010: 63). The TechnoDolly, as a cutting-edge tool for motion control, is no exception to the rule. What it teaches, at least in the hands of the *Orphan Black* cast and crew, is that *the TechnoDolly is us*. In its most intimate and engineered scenes, the programme shows us a calm alliance of human and machine, of a technical actor and an organic apparatus.

The TechnoDolly Escapes Her Programming

> Thus actors, particularly female actors, created the beginnings of what we now call psychologically real acting [...].
>
> (Cima 1996: 35)

At the same time, the TechnoDolly might be considered an emblem (however extreme) of all realist acting since the mid-nineteenth century. The word 'realist' may sound strange when applied to the acting of a show like *Orphan Black*. In early episodes (and off and on throughout its run) *Orphan Black* is ruled by the camped-up conventions of familiar genres – from the hard-boiled procedural to the slow-simmering soap, from the dead-eyed slasher film to the misty-eyed family drama. But this is only where the show begins, not where it's heading. As Loofbourow (2015) suggests in her essay, 'The many faces of Tatiana Maslany', *Orphan Black* conjures these genre conventions only so that Maslany (and her characters, and her audience) can defy them eventually. The show asks,

> What does the exact same woman look like if you grow her in the petri dish of *Desperate Housewives* or on a horror-film set in Eastern Europe? What about a police procedural?

> The result is a revelation: Instead of each archetype existing as the lone female character in her respective universe, these normally isolated tropes find one another, band together and seek to liberate themselves from the evil system that created them.

Within the world of *Orphan Black*, this 'evil system' is a worldwide cloning conspiracy, but in our world, it's a parallel order: the 'evil system' of screenwriting and casting, which relentlessly 'clones' the same women, the same female archetypes, from one project to the next. So, as Maslany's many characters collide, so do their genres, their styles of acting, their ways of being and the scripts they always follow Alison, the desperate housewife, breaks bad; Helena, the sadistic assassin, finds a family and falls in love. No one stays in her genre for long, and no one stays in her place.

A famous strain of realist acting began in the late nineteenth century with just this sort of woman: one who refuses the role assigned to her by fate, by society, by her husband and most of all by a playwright. These women (I think especially of Nora Helmer, the lead character in Henrik Ibsen's play *A Doll's House* [1879]) start off thinking of themselves as the heroines of melodrama, but leave the stage as modern, realist women. This kind of play required a whole new sort of acting, one where women first inhabited, then transcended conventional roles. They would, in fact, play many roles – not one after the other, but entwined in one play, in each moment they spent onstage. *A Doll's House*, says theatre historian Gay Cima, requires at least two different Noras at once:

> [...] the actor playing Nora must portray not only the Nora of the realistic play – a woman strong enough to forge her [father's] signature, to work secretly to repay a debt, and to walk out on her husband [Torvald] – but also the Nora of the melodrama unfolding in the character's mind, the birdlike, game-playing Nora who has heroically 'saved' Torvald, and who envisions herself 'tied to the tracks' as Torvald, in turn, rescues her heroically. The actor must reveal where Nora thinks she is headed as well as where she actually is headed.
>
> (Cima 1996: 44–45)

Nora Helmer is woman and doll at once, autonomous creature and automaton – shallow damsel and deep, realist character. The prototype of the realist character, then, is a woman who dares to escape her programming – the only human, so to speak, in a world of robots. This anyway is the joke behind Elizabeth Meriwether's *Heddatron* (2006), a burlesque on Ibsen in which a woman – pregnant, depressed and too often alone – is kidnapped by robots, taken to a distant jungle, and forced to star in an all-robot production of Ibsen's *Hedda Gabler* (1891). I sometimes think of Tatiana Maslany – a woman who, some days, acts with tennis balls and TechnoDollies only – and I know she belongs with Nora Helmer, with Hedda Gabler and with *Heddatron*'s desperate abductee.

But if the prototype of realist character is a de-programmed woman, she doesn't lie beyond all programmes; she simply engages them in novel ways. What Elizabeth Robins,

England's first Hedda Gabler, showed audiences was 'the simultaneity of the melodramatic and the "real"' (Cima 1996: 45). This new acting style, in other words, was defined – not *against* genre – but through a slippery engagement with several of them. By playing around with multiple genres, the actor can 'control her audience's shifting sympathies and make them uneasy with any attempt to pigeonhole her character' (Cima 1996: 40). This character never escapes into that unspecified space beyond artifice, beyond genre, beyond programme or protocol. Instead, she stays 'sensitive, responsive, mobile – like a well-oiled and regulated machine' as genres mix and collide. She takes what she needs when she needs it, confirming or denying each genre convention as the play and her performance requires.

Audiences say that what they see is 'realistic', but what they mean is that it's *more realistic* than the kind of performance that sticks to one genre. If you look at two similar but not identical images, routing a different one into each of your eyes, a strange thing will happen. Any difference between them, however small, will start to wobble or shimmer until your brain resolves the conflict. This is how the book historian Randall McLeod, using a device of his own invention, examines the tiniest variations in different printings of old books (McLeod 2011). It's also how the stereoscope turns a pair of two-dimensional photographs into an illusory three-dimensional vista. If the actress playing Nora or Hedda was doing things right, those early realist performances would, at first, wobble and shimmer with a kind of binocular conflict. This would give her audience a strange sensation, which they would come to label 'realism'. But then these shimmering portraits would slowly smooth out and sink down into a three-dimensional vista: the illusory depth of a realist character.

When you watch *Orphan Black*, each of the clones Maslany plays should shimmer at first with binocular conflict – none more so than Alison Hendrix, suburban housewife and latter-day Nora Helmer – but then they'll all settle down into realist depth. *Orphan Black*, though, has its own special ways of reviving and prolonging this kind of effect: by setting ten different characters on the body of one woman, it keeps the machinery of realism in view even as audiences start to sink into the illusion. When Sarah touches Alison, embraces Cosima, or tackles Helena, each character starts to ripple and shimmer again. And when, at the end of Season 2, they start dancing together in one long TechnoDolly shot, you can experience this effect in its purest form. More than that, this sequence literally makes a spectacle out of the work Maslany does off-camera to build and sustain character. Remember those playlists she built for each clone – remember her habit of dancing her way into character? Here we see it on screen – not just the process of one actress, not just an emblem of all realist characterization, but also finally what the theorist Erin Manning calls 'individuation's dance'.

> Think technique not as an add-on to a pre-existing body-form but as a process of bodying. Think technique as an in-forming of a mutating body. And then think that body as a field of relations rather than a stability, a force taking-form rather than simply a form.
>
> (Manning 2013: 31)

This is precisely what *Orphan Black* encourages us to do. The characters it shows us are, to quote the title of Manning's book, 'always more than one'. They are clones and individuals – played by actresses and their doubles. They are 'very technical' women who are learning to play with 'organic' TechnoDollies.

Conclusion

The technologized body, ghosts of others in our innermost selves, the elaborate codes behind the 'natural', the 'organic' – these are some of *Orphan Black*'s most obvious themes as a show about women and cloning. But they are also, I hope to have shown, simple descriptions of the work this show demanded of its actors. Following Haraway and Barad, I consider this performance mode feminist, even apart from the show's feminist message. These actors and their characters are all cyborgs – not because they've been dehumanized, but because human and machine are intra-acting. The work they do is not their own – not because it has been stolen, but because they were 'always more than one'.

None of this should sound outlandish: it's just a description of life, though one written in six-foot-tall, flashing letters. We are all cyborg creatures, shaped by our lively machines, informed deeply by the members of our own 'ensemble'. We are cobbled together from scraps, few of which belong to us – none of which will make us 'natural' or 'original' as, in her more murderous days, Helena yearns to be. After you've watched a few seasons of *Orphan Black*, try going back to Maslany's recorded interviews or her Q&As at Comic Con. All of a sudden, you'll see a smile spread across her face – Sarah's smile, which will burst out, oddly enough, into Cosima's laugh. Identity itself will wobble and shimmer for a moment, but then sink down into the usual perspective: the realist illusion of a deep individual.

Notes

1 This essay, too, is an 'ensemble performance'. Thanks to Jacob Gallagher-Ross, who invited me to SUNY-Buffalo to make an early version of this argument – and to Shonni Enelow, my co-presenter that day. Both of them have shaped my thinking a lot, and not only in this essay. Thanks also to Justin Rawlins for his advice years ago on how to enter scholarly conversations on TV acting. Finally, I salute my talented students at Amherst College who, in a seminar on 'Technology & Performance', were an early audience for these ideas and a great group of scholarly co-performers.

2 None of this is quite complete or correct, of course. For two recent, scholarly takes on the Method, see Enelow (2015) and Gallagher-Ross (2015).

3 For a brilliant analysis of the Method's focus on privacy, see Gallagher-Ross (2015: 294 ff).

4 For more on 'technical' and 'organic' acting styles, see Bell's essay in this book.

5 In contemporary Hollywood, the vast majority of animators are male, and the only majority-female sectors of film crews are those who do historically feminized labour – e.g. make-up

crews. Some studies suggest that the polarization of film labour by gender has only gotten worse. According to Stephen Follows (2014), in the years between 1993 and 2013 the percentage of make-up artists who were women increased by 7.1 points, while the percentage of animators who were women fell by 18.4 points. These were, respectively, the largest gains and largest losses found in any one sector of the industry (see Figure 5.3 in Follows's report). In other words, the gendered connotations of the phrase 'digital make-up' would never have rung out so clearly as when Serkis lobbed the slur at animators in 2014.

Animating Cloning: Special Effects and Mediated Bodies in *Orphan Black* and *Jurassic Park*

Simon Porzak

Orphan Black tells a rather unapologetically sci-fi story, tracing the conflicts between a group of bioengineered clones and the shadowy corporations and futurist movements that seek to control them. Yet critical celebrations of the series often downplay this science-fiction content and privilege another centre of interest: the intersecting lives of those clones, brought to spectacular life by the actress Tatiana Maslany. One typical review describes the show as an 'only semi-science-fiction exploration' of identity that becomes 'a quirky showcase for a seriously talented actress' (McNamara 2014). Another contrasts *Orphan Black* to the 'very different' preceding programme on BBC America, the august science-fiction *Doctor Who*; it concludes that 'slotting in the new show on Saturday looks like more of a convenience for the network than anything else; *Orphan Black* is not truly a *Who* companion piece, and its [television] brothers and sisters are decidedly earthbound' (Ryan 2013). Other critics explain the show's sci-fi as an excuse to portray the real biomedical experience of contemporary women (Murphy 2016). *The New Yorker*'s Jill Lepore extends this thesis to female experience writ large, proclaiming that 'what has been so deliriously fun about the first two seasons of "Orphan Black" isn't the fiction of the science; it's the fact of the women' (Lepore 2015).

These readings, many of them seeking to highlight the feminist importance of the show's dramatic and intellectual work, frame their critical generosity in terms of an indulgent ignorance about *Orphan Black*'s sci-fi tropes. And yet, without the science-fiction machinery of the show – the fictional technology allowing for so many clones to confront so many 'deliriously fun' challenges – there would be no showcase for Maslany's talent, no exploration of identity and medicalized violence, no women at all. The show, therefore, must trace a more complicated relationship between the aesthetics of its science-fiction and the fact of its feminine interest, even as it seems to want this connection to appear as invisible as possible.

Such critical oversight is easily forgivable. After all, the creators of the show themselves foreground the series' 'gritty realism' 'in spite of [its] sci-fi premise' (Mele 2013); during the publicity tour for its first season, one of its stars described it as 'so different than anything I've ever come across or seen. I don't think it relies on gimmicks or tricks or anything like that at all' (Mele 2013). Certainly, *Orphan Black* stages itself so as to look as different as possible from those shows that rely on sci-fi 'gimmicks or tricks'. Take, for example, the works of J. J. Abrams, whose *Alias* (2001–06), *Lost* (2004–10) and *Fringe* (2008–13) helped define the contemporary aesthetics of science-fiction television. Like *Orphan Black*, these shows sell familiar genre-TV plots (spy thriller, adventure and police procedural respectively) by dressing them up with sci-fi twists; *Orphan Black* often presents clone-spiked versions of corporate espionage dramas (the good clone Cosima tries to take down the evil Dyad corporation from the inside), noirish cop shows (the clone Sarah poses as crooked detective clone Beth) and even blackly humorous family sitcoms (the clone Alison ruthlessly schemes her way through placid suburban soccer games and community theatre productions).

Abrams's series, though, follow through on their sci-fi promise: the gimmick always arrives, in the form of a Renaissance bioweapon on *Alias*, a polar bear on a desert island in *Lost* or a genetically-engineered monstrosity in *Fringe*. And you can't miss it, since these shows make their sci-fi elements look so obviously sci-fi: they frame them jarringly against a mundane backdrop or accent them with musical cues to make them all the more unmissable. This effect of contrast extends to the clash between the straight filming of the framing scene and the sometimes-rickety computer graphics or make-up effects used to produce these exceptional elements.

Even when the science-fiction element does not occupy a central position in the frame – as in Matt Reeves's 2008 film *Cloverfield*, produced by Abrams, and where the monster mostly scuttles around the margins of the image – it remains the focus of our desire as science-fiction viewers, the thing we want to see, the object promised by the genre. Albert LaValley defines audio-visual sci-fi in terms of this movement towards 'the manifestations of ever more spectacular special effects', with this 'formal rhythm of the science fiction film' defining the experience of the sci-fi viewer, who cares less about plot and character than the alternations between realist and special-effects-driven filmmaking (LaValley 1985: 148). With very few exceptions, *Orphan Black* elides or minimizes such obvious moments of sci-fi spectacle, eliminating them from most episodes instead of presenting them, Abrams-style, as the very climax of each episode's interest. Winkingly, the most glaring exceptions to this rule all seem to involve graphic damage to the eye: Rachel gets shot in the eye with a MacGyvered pencil gun, Rachel scoops out her own bionic eye. Although such scenes end in an obvious climax of near-futuristic body horror, they also seem to comment on how such horror normally lies, as it were, in the series' blind spot. Like the maggot-bot that can only be seen through Sarah's cheek in the light of rear-projection, when such sci-fi spectacle appears, the series carefully marks it as belonging to another way of looking at the same subject. Still, fans keep looking for such effects, and *Orphan Black* keeps frustrating their expectations; the show's make-up artist must often respond to inquiries about the prosthetics he uses to build Maslany's cloned faces by explaining, quite simply, that he doesn't (Loofbourow 2015).

Commenting on *Cloverfield*, Daniel North observes that, in the case of the science-fiction viewer, our readerly demand to 'experience the spectacle' of science-fiction 'is closely linked with our fascination with the novel visual effects technologies often tasked with animating them, creating interdependence between cinematic spectacle and cinematic technology' (2010: 74). Certainly, *Orphan Black* solicits this desire in its fans, as witnessed by their questions about Maslany's facial prosthetics, or the DVD features that breathlessly explore Maslany's relationship with her acting double or the complicated camera rigs used to generate the show's seamless images. Yet in contrast to *Cloverfield*, where the trace of the sci-fi monster is always barely present in the image, *Orphan Black* elides recognizable special effects in such a way that other readers can simply view it as not, or not really, science-fiction.

Perhaps this choice represents a canny marketing strategy on the part of the show's producers, who seek a wider or more prestigious audience beyond the ghettoized limits of genre TV. Or perhaps it stems from a general problem with clones – since they look exactly like the things they replicate, they cannot stand out from the image in the spectacularly monstrous manner of Abrams's uncanny beasts. By exploring the way *Orphan Black* makes its science-fiction content simultaneously visible and invisible, and in particular by locating its visual rhetoric within a genealogy of visualizations of cloning, I'd like to suggest that the show offers a new understanding of how science and fiction intersect, framing these clashes of genre and medium as a series of unmasterable clonings.

When Clones Meet: *Orphan Black* and *Jurassic Park*

So, what is traditional science-fiction supposed to look like? As an example, let's look at the first extended encounter between palaeontologists Alan Grant and Ellie Sattler (Sam Neil and Laura Dern) and the reanimated dinosaurs of 1993's blockbuster *Jurassic Park*. The two have been summoned to a remote island by John Hammond (Richard Attenborough), the CEO of the bioengineering firm InGen, to verify the safety of his new project: an amusement park featuring live dinosaurs cloned from dinosaur blood recovered from mosquitoes trapped in amber. As the scientists travel from their helicopter to the visitor centre, they interrupt their ride to gaze in awe at the prehistoric replicas grazing on a bucolic plain. As the score swells and builds to its epic leitmotif, we follow Alan, and then Ellie, as they rip off their sunglasses in amazement, the film teasing us by building our desire to see what they see. Then, in a reverse-shot, the camera pans from their backs as they stand in the Jeep and up over the grazing Brachiosaurus. 'It's… it's a dinosaur', Alan stammers, and the camera zooms in on his eyes.

This sequence clearly establishes the dinosaurs as objects of sublime wonder (soon to turn to terror as the animals escape their pens and retake the island). At the same time, it establishes the dinosaurs' *images* as objects of a similar wonder. In its day, *Jurassic Park* represented 'the ultimate triumph of computer simulation', thanks to its 'seamless integration of film footage of real scenes with computer simulated objects' (Manovich 1995: 14–15). When the camera cuts from the eyes of Alan and Ellie to the dinosaurs they see, then, it insists on the perfect symmetry between their vision and the vision of the film spectator: they see the amazing technology used by InGen to make the dinosaurs, and we see the amazing technology used by Industrial Light & Magic to make the dinosaurs' images, to make their computer animations look *just like* film stock. We gape both at the cloning tech within the fiction, the method developed by InGen to bring dinosaurs back to life, and the cloning tech outside the fiction, the method developed by ILM to bring InGen's reanimation to cinematic life – and since the same image wows us in both ways, it's hard to tell where the one special effect begins and the other ends.

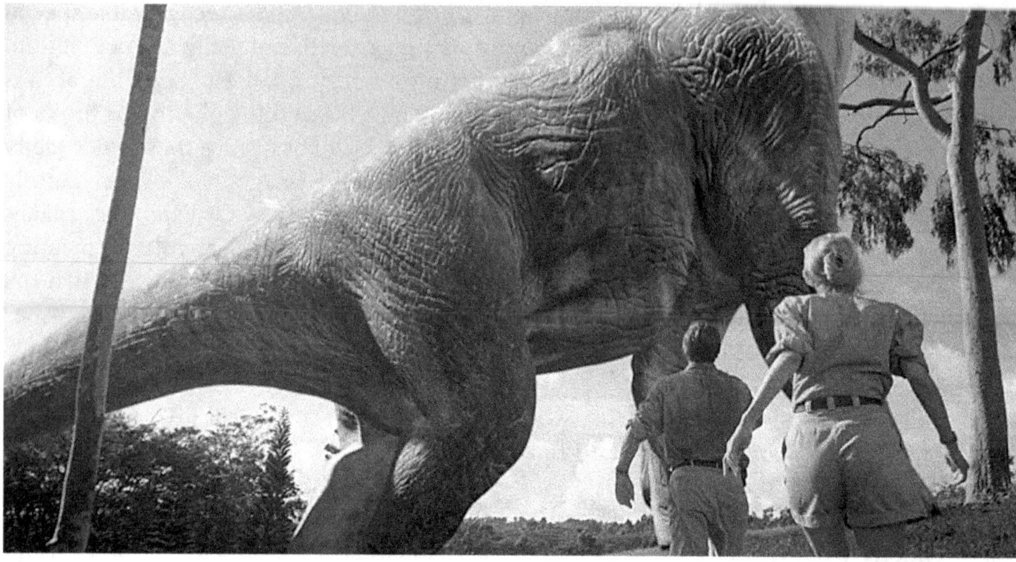

Figure 5: The dinosaur as sublime object of wonder, too large to fit into the frame, dwarfing the human observer.

Like *Orphan Black*, *Jurassic Park* has already begun with a teaser scene in which clones are ever-so-briefly glimpsed before their death (a velociraptor escapes its cage during transport and is shot in the ensuing melee; Sarah sees Beth before she throws herself in front of a train and Katja before Helena assassinates her). But *Orphan Black*'s first visual depiction of a sustained encounter between clones (Figure 6) looks nowhere as spectacular as *Jurassic Park*'s. Following a mysterious phone call, Sarah has traced the mystery surrounding her adopted identity to a suburban parking lot, where she encounters her doppelganger. If she reacts first with shock – 'Holy shit!', she exclaims (the show plays it off for laughs) – 'A soccer mom', she continues, disgusted at the difference between her and her twin.

Similarly, as the camera follows her into the portable building where said soccer mom is preparing snacks, the show continues to de-emphasize and diffuse its spectacle. As if we've wandered off-set and found the undecorated backside of its glossy theatrical flats, the shoddiness of the plywood panelling highlights the cheapness of the very non-special effects technology used to produce the image (the technique of double-exposing and matting dates back to the dawn of motion pictures). At the same time, the cloning technology escapes from the clearly delineated plane that it occupied in *Jurassic Park* and hides itself within the bodies of the two women; the scene does not clearly direct the viewer towards a site of sublime wonder.

If the technology does become visible, it's in the knife that Alison, the soccer mom in question, fidgets with anxiously throughout the scene. The knife signals not only Alison's aggressiveness towards the suspect Sarah, but the technology that was used to splice together

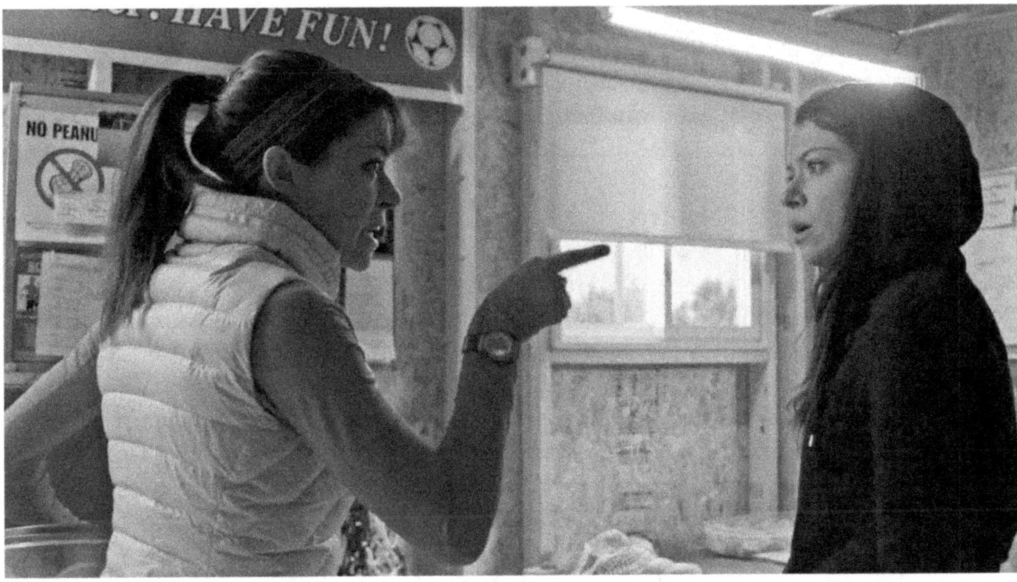

Figure 6: Clones first meet in *Orphan Black*.

two takes of Maslany into one seemingly seamless image – as well as the technology that was used to split one genome into a multiplicity of clones. Instead of painting a digital effect onto or into the backdrop of a non-fantastic image, as in the garden framing the Brachiosaurus, the sequence dissolves the boundary between the real shot and the special effect. Since the sequence would be invisible *without* the splicing technology that occurs in its every frame, the special effect here is what allows us to see anything at all. As such, it's transparent; the special effects in *Orphan Black* aim to make themselves completely invisible, so that there's nothing 'special' happening at all (unlike the appearance of a giant dinosaur). The digital cloning of Maslany is what allows normal things to happen on camera (people talking, arguing, dancing). Like a lens, it's transparent and allows for there to be vision in the first place. Without marking any passage from one visual order into another, as Alan does by removing his sunglasses to signal a shift in his perspective, *Orphan Black* codes its special effects into the basal functioning of our eyes.

Jurassic Park and *Orphan Black* share a lot of DNA. They both centre on an ethically ambiguous corporation's attempts to control a cloning experiment, for motives that vacillate between scientific altruism and an economic bottom line. More interestingly, they both define clones, and their capacity to escape the control of scientific programming, as female. *Jurassic Park*'s scientists reanimate only female dinosaurs in order to control their population, and nature triumphs over man's reason when the dinosaurs are able to spontaneously change their sex and procreate (they hijacked some of the amphibian genes used to restore their genome); in *Orphan Black*, Sarah and Helena's unique fertility as the only clones able to bear children make them of tantamount importance to the ongoing experiment. Yet these similar themes appear not to translate into the final aesthetics of their visual presentation. So how might we draw the taxonomic connection between their visual rhetorics, the one as interested in maximizing its blockbuster spectacle as the other is in downplaying its biological sci-fi?

On feminist geek blog The Mary Sue, Isabella Kapur attempts to sever any tie between the two styles of sci-fi. She celebrates *Orphan Black* for focusing 'on biology for its feats of futuristic fiction, not silicon or metal. In fact, the show's clone premise is what makes it particularly modern. While we see a lot of sci-fi stories focusing on war, the consumption of our world by technology, or the exploration of aliens and other planets, *Orphan Black* focuses on humans, here and now, in a fantastic way' (Kapur 2013). This reading would link a turn to the biological interior of the body to an aesthetic rejection of the spectacularly external, locating the 'modernity' of *Orphan Black* in such a refusal of the obvious visualizability of the alien. (Of course, Ridley Scott's *Alien* franchise has recently revealed itself to be entirely about genetic engineering.) *Jurassic Park* was able to make clones look spectacularly monstrous, so what could be 'fantastic' about *Orphan Black*?

The word 'fantastic' crops up a lot in coverage of *Orphan Black*. The term harkens back to a time before science-fiction, to the eighteenth-century genre epitomized by E. T. A. Hoffmann (1776–1822) that would later branch into the more-defined subspecies of science-fiction, horror and fantasy. For literary scholars, the fantastic inexorably recalls the 1919 study of Hoffmann by Sigmund Freud (1856–1939), 'The uncanny', in which he defines the fantastic genre by its capacity to generate a feeling of 'uncanniness' in its reader, an affect that accompanies the re-emergence of a repressed memory, wish or other psychic content (Freud 2003). For Freud, 'silicon or metal' automatons like Olympia in Hoffmann's story 'The Sandman' are uncanny, in that they confront us with the inescapable biological programming of instinct that drives us, like robots in the face of their coding. So are doppelgangers, the duplicated selves that appear in one of the first films to employ the suturing technique perfected by *Orphan Black*, 1913's *The Student of Prague* – which was analysed extensively in a paper by Otto Rank (*The Double*, 1914) that Freud also cites in his study. So, at this fantastic time, the strands of technological and biological sci-fi had not yet diverged, and they both appeared not at all alien, but as disquietingly familiar, as something that looks normal but isn't.

But I want to pursue Kapur's suggestion that '*Orphan Black* focuses on humans in a fantastic way'. She means to suggest that, on the level of story, the show presents 'fantastic' humans (not-quite-humans, clones), yet her grammar instead argues that the show's *way of focusing*, its visuality, is fantastic. Consequently, we must take a second look at *Orphan Black*'s way of looking and the way it looks. This will entail a certain critical splicing of observations about film, digital cinema and television, trespassing on one of the most major boundary-lines in media theory – the strict genealogical separation of film and television. Yet just as the 'fantastic' reunites robots and clones, *Orphan Black* finds its evolutionary niche in our new-media ecologies: streaming between phones, computers and television

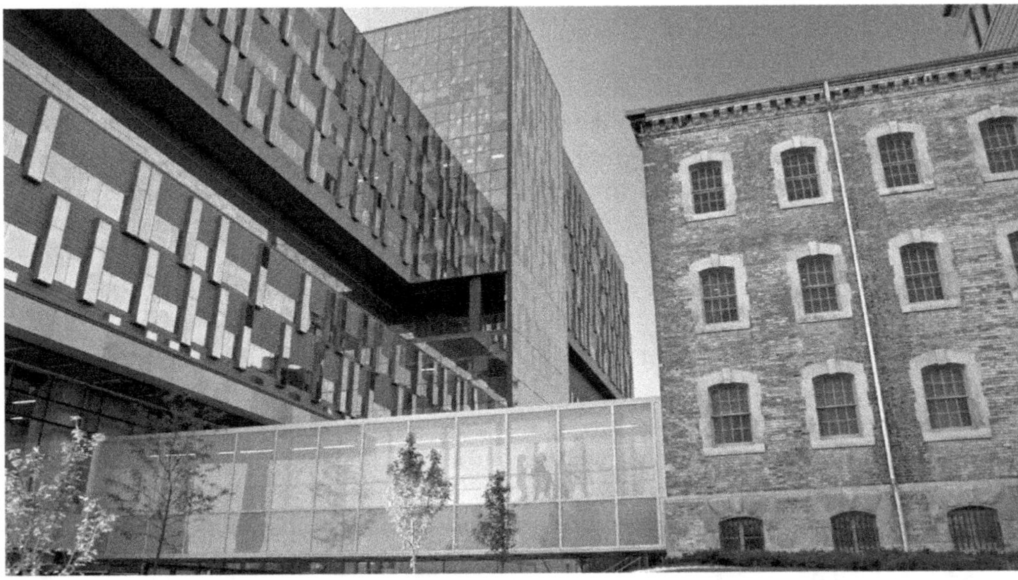

Figure 7: Silhouettes in Dyad's glass-screened corridor.

screens attached to cable boxes or gaming consoles, the show gained its audience by being spliced into endless images and GIFs for online discussion. Possessing the eerie transportability of digital media, *Orphan Black*'s images recall David Bolter's and Richard Grusin's thesis on new media as remediation, as an accelerated process of media mediating other media in order to present a sense of immediacy by increasing the multiplicity of their mediated foundations, 'to erase [their] media in the very act of multiplying them' (Bolter and Grusin 1999: 5). More importantly, *Orphan Black*'s aesthetics insist on mediation as a kind of perpetual cloning linking film and television, generating its fantastic world of perpetual (re)animation.

But it's hard to see the way something feels or looks; you can only catch it in offhand moments. You find it in the second episode of the second season, when Cosima is being courted by the 'Neolutionist' bio-futurist Dr Leekie and his magnetic assistant Delphine. They want to recruit her into the Dyad corporation, the force driving the clone experiments; they take her on a tour of her new laboratory in order to convince her that she can do more good working within the organization than against it.

The sequence begins with an establishing shot of Dyad that gets Xeroxed into episode after episode – but this time, the trio is visible through the building's glass corridor, their silhouettes like shadow-puppets as they move from the economic to the scientific building. They're talking about something, vaccines maybe, or whether Dyad is good or evil. In the lab, Cosima is underwhelmed at first but Dr Leekie assures her that her 'opportunities will be endless': 'Just last month', he continues, 'I emailed a vaccine to a colleague in Delhi who then re-created it on a 3D biological printer!' Cosima blinks: 'That is', she mutters, her head shaking, 'completely amazing'.

Like Alan and Ellie in *Jurassic Park*, Cosima's reaction mirrors and models that of the viewer. Blinking incredulously, she attempts to determine whether what she's seeing is real or fiction; the image conjured up by Leekie's jumble of signifiers seems at first like some sort of hallucination – but is it? Could you actually email a vaccine halfway around the world and then 3D print it? All of those words feel meaningful individually, but put together they become impossible to evaluate. Science-fiction might be defined as the genre that features imaginary technologies – from Jules Verne's *Nautilus* (in *20,000 Leagues Under the Sea*, 1870) to the titular *Battlestar Galactica* (2004–09). But here, I simply do not know whether this is science-fiction technology or real technology. The 'CSI effect', notoriously, has made criminal trials in America quite difficult, since today's jurors expect police investigators to employ the techniques and machinery they've seen on *CSI* (another stealth-sci-fi TV programme that doesn't ever visibly announce the near-futureness of its technological marvels), even though these techniques sometimes don't exist. But unlike these jurors, whose failure to recognize the science-fictionality of *CSI* proves their certainty – they know they're right and that these techniques exist; they know that *CSI* employs a strict realist aesthetic – I cannot figure out how to *take* Dr Leekie's statement. Am I supposed to be amazed by the near-future technology he showcases, or instead by the fact that it's actually technology of the present? Is this plot device a sci-fi spectacle or part of *Orphan Black*'s 'gritty realism'?

Similarly, the establishing shot of the hallway confounds explication. To produce the image of Cosima, Delphine and Leekie walking down the hallway, the producers could have filmed them walking down the hallway, or used lookalike doubles. Or were they, or others like them, filmed elsewhere and then composited in? Or are the silhouettes digital animations painted onto the outside of the corridor's glass screen? Have these images been glued together digitally, or by the real-world techniques of light and shadow that project images spontaneously onto other images? As with the biological printer, it's impossible to tell whether this effect comes from an uncertain realism or an uncertain reality.

'An uncanny effect often arises when the boundary between fantasy and reality is blurred', mentions Freud near the end of his study (Freud 2003: 150). What's special about *Orphan Black*'s special effects is that you can't tell where they start and where the non-special effects end, all the more so in that they emphasize how fantastic the real world already is. This returns to the definition of the uncanny as 'intellectual uncertainty', put forward by Freud's colleague Ernst Jentsch (1867–1919) and later re-appropriated by Tzvetan Todorov (1939–2017) as the defining feature of the genre he called 'fantastic', which Freud had hoped to elide in his essay. Jentsch considers the example of 'a wild man' who 'has his first sight of a locomotive or a steamboat', concluding that the uncanny terror that seizes him emerges from his incapability to decide whether said technology was the work of man or the work of nature, reality or a fantastic chimera (Jentsch 1997: 11). Reminiscent of Arthur C. Clarke's dictum that '[a]ny sufficiently advanced technology is indistinguishable from magic', this definition destabilizes an understanding of science fiction as inhering in any recognizably futuristic or projective object. Instead, as *Orphan Black* shows, science-fiction becomes the most 'fantastic' when the viewer cannot tell what is or is not science-fiction, so that an uncertainty about the limits of reality and fiction, science and fantasy, forms the heart of the genre's unique effect. Certainly, *Orphan Black* elides the obvious rhythmic alternation between realism and special-effect that LaValley described as the central pleasure of the sci-fi film, but in so doing it generalizes and heightens the climactic 'interface between the two worlds', the promised collapse between present and future, that forms the fantastic promise of the genre (LaValley 1985: 145).

So where does this intensification of a collapse between reality and projected reality place *Orphan Black* in the genealogy of sci-fi? Jean Baudrillard, reviewing the novels of J. G. Ballard, observed that in our virtualized age, where the easy distinction between Real and Imaginary has collapsed, science-fiction needs only to precisely mirror everyday life in order to capture the technologized forms of indistinction and collapse that now define our 'reality' (Baudrillard 2006). *Orphan Black* has spawned multiple comments and even books about how 'realistic' its science is; *The New Yorker*'s Jill Lepore celebrates that realism as proof of the show's anti-sci-fi nature. But simply because *Orphan Black*'s world 'is *our* world, nothing in it is "invented"', as Baudrillard observes about Ballard's *Crash* (1973), that need not make it any less *fictional*, since the utopian space of early science-fiction, which once held up a mirror to our world from the distance of fantasy, has been absorbed into our reality (2006: 125). Baudrillard expresses this in the language of the double, which once operated its replication from a disquietingly close distance: 'the double has disappeared, there is no longer a double, one is always already in the other world, which is no longer an other, without a mirror, a projection, or a utopia that can reflect it' (2006: 125). Science-fiction, then, would be a medium of cloning – a kind of absolute realism (due to perfect reproduction) that would nevertheless capture the strange irreality of our technologically displaced present.

Although Baudrillard celebrates the negative, implosive power of this kind of science-fiction *of* cloning, he elsewhere rejects real-world projects of cloning and genetic engineering. Indeed, Baudrillard might find *Orphan Black*'s emphasis on bio-techno-futurism, in contrast to the decidedly everyday automobiles and freeway overpasses of Ballard, decidedly regressive. For Baudrillard, the clone and cloning technology ought to remain taboo, unimagined, because the clone 'realizes' the fantasy of the double, materializing something that must remain immaterial. The double isn't just any fantasy; instead, it's the very matrix of fantasy, the image that opens up the possibility of dreaming of difference, because it forms the structure allowing us to dream that things could be just like they are, except... Consequently, the double constitutes the very *medium* of fiction, its obvious, visible irreality proof of the capacity of fantasy and desire to diverge from reality; when we reproduce the double in the flesh, we foreclose this space for differentiation and metamorphosis. In *The Vital Illusion*, Baudrillard frames this argument in cybernetic and evolutionary terms, discussing the emergence of sexed life as the advent of 'a duality that puts an end to perpetual indivision and successive iterations of the same' (2001: 9); cloning, by rewriting the individual human as an informational sequence that can be perfectly iterated, returns the strange non-life of asexual reproduction to the human species, and turns death into pure 'virtual reality', 'an option or changeable setting in the living being's operating system' (2001: 11). Without a space for death, the cessation of being as we know it, we can neither imagine nor propel ourselves into a transformed future. From this standpoint, *Orphan Black* and *Jurassic Park* would together constitute two modes of a simulacral sci-fi that, as opposed to the negative power of *Crash*, would seek to reinforce the deathly force of our technologies of simulation.

But can we so quickly counterpoise these two forces of simulation? Baudrillard acknowledges the facility of this opposition in a footnote to a similar chapter in *Simulacra and Simulation*, where he muses that no perfect cloning could ever be possible, because of an 'eccentric distortion' inherent to materiality as such (2006: 102). In other words, because even pure genetic information is carried by a physical genome, 'millions of interferences', as Baudrillard says, will necessarily intervene, 'demonstrating the radical impossibility of mastering a process simply by mastering information and the code' (2006: 103). This self-subverting auto-differentiation inherent to all information, which must – on some infra-level – always be material *qua* mediated in a physical body, is what allows asexual reproduction to produce evolutionary mutations, and can never be eliminated in cloning. Baudrillard attacks 'cloning' because he believes it realizes the double and as such removes difference from fantasy; perhaps instead, cloning – in its stubborn refusal to remain predictably or knowably real – reveals the originary and radical irreality of the body, the differentiating fracture within the Real. *Orphan Black*'s science-fiction of cloning need not add anything to the body's putative integrity in order to reveal the ways any body interferes with its own imaginary stability. In other words, by removing the obvious contrast between special-effects shot and realist framing image, *Orphan Black* reneges on the generic contract to present the novel future 'as other, as different', removing the recognizable 'leap from the world we are in' (LaValley 1985: 145). But in so doing, it shows us that this leap may already be taking place, invisibly, within the very fabric of that world. Perhaps *Orphan Black*'s pervasive paranoia is another form of the heightened attentiveness to the elusive, transparent self-subversiveness of Baudrillard's bodily Real.

The philosopher Martin Heidegger (1889–1976) had a name for the force of self-subversion inherent to materiality; he called it *time*, the process by which something can change and so become a part of a future. For Charles Darwin (1809–82), this force was *evolution*, the process by which the ordinary course of an organism's life could open onto radical new modes and structures of existence and so give a temporal arc to the branchings of natural history. This rhythm defined the sci-fi genre for LaValley, but isn't it also precisely the force of time that appears in cinema, as the effect special to cinema as a medium? Film depends, as Friedrich Kittler emphasizes, on spatiotemporal disunity, the 'reels, cuts and splices' that allow for special effects such as those used in the fantastic cinema (1999: 115). Similarly, Christian Metz, attempting to define the various rhetorics of trick-shot or *trucage* at work in the cinematic image, concludes that 'montage itself [the patterning of images that gives the cinema its temporal dimension] is already a perpetual *trucage*, without being reduced to the *false*' (1977: 672, original emphasis). For these reasons, Lev Manovich argues that film, since it is 'based on sampling – the sampling of time', 'prepares us for digital media' and their techniques of reducing a scan to a binary series (1996: 4), a procedure that collapses the distance between digital film and analogue/chemical film, as the DV format of digital video simply samples the sampling inherent to film. *Jurassic Park* does not add anything new to the cinematic experience by splicing in computer-generated images; *Orphan Black*, then, does not subtract anything from the visual rhetoric of science-fiction by eliminating the distinction between the real image and the imaginary image, but instead reveals the fundamental indifference, the fantastic equation of the two, at the heart of sci-fi aesthetics.

But which comes first – film, or the sampling it employs to fantastic ends? Kittler's pursuit of the origin of the 'trick film' epitomized by the special effect of the double ends in frustration. He can find 'no datable origin' to such tricks and thus concludes, like Metz, that 'the making of films is in principle nothing but cutting and splicing: the chopping up of continuous motion, or history, before the lens'; unlike Metz, Kittler sees this process of splicing that ends up consuming its own productive matrix as bringing an end to film history and history itself (1999: 117). There is no *first* trick film – only doubles who deconstruct the very possibility of an originality about which we could be intellectually certain – and this fact feels uncanny, a bit like when we learn that the first televisual experiments predate the invention of film cameras and fax machines. In response to this uncanny fear, Kittler accuses cinema of imposing a violent, serial cutting and splicing onto the Real, identifying its violence as external to a stable underlying reality. But the science-fictions of cloning, such as *Orphan Black* and *Jurassic Park*, instead suggest that the materiality of genetics, the very matrix of the bodily Real, belies such a fantasy of an originary continuity of motion that would then be split by the trickery of technological special effects. To see how, we must first turn to the alternate history of cinematic media proposed by *Jurassic Park*, and then trace its ramifications through the uncanny animation of *Orphan Black*'s clones.

Animation, Image Splicing and Cinematic Hybridization

After witnessing the dinosaur herds, the palaeontologist heroes of *Jurassic Park* have one question: 'How'd you do this?' If Hammond whispers 'I'll show you' in response, he explains not by showing them the cloning lab, but by showing them the mock-up of the video that will answer the question for the gaping masses at the island's visitor centre. Hammond welcomes them to an amusement park 'incorporating all the latest technologies', but his welcome video is charmingly low-tech: it relies again on the same double-exposed trick shots and ambiguous projections as we saw in *Orphan Black*, and adds in some Saturday-morning-cartoon-level animation to boot. The edutainment centres on the conceit that the real Hammond explains to a grainy film-stock Hammond how he's been cloned (Figure 8), visualizing this by taking a drop of blood from his image that becomes the genie-like Mr DNA (Figure 9).

Mr DNA fills in the audience(s) on the conceit of Hammond's film: Jurassic Park's scientists have retrieved dinosaur blood from the bodies of mosquitoes trapped in amber. Since some

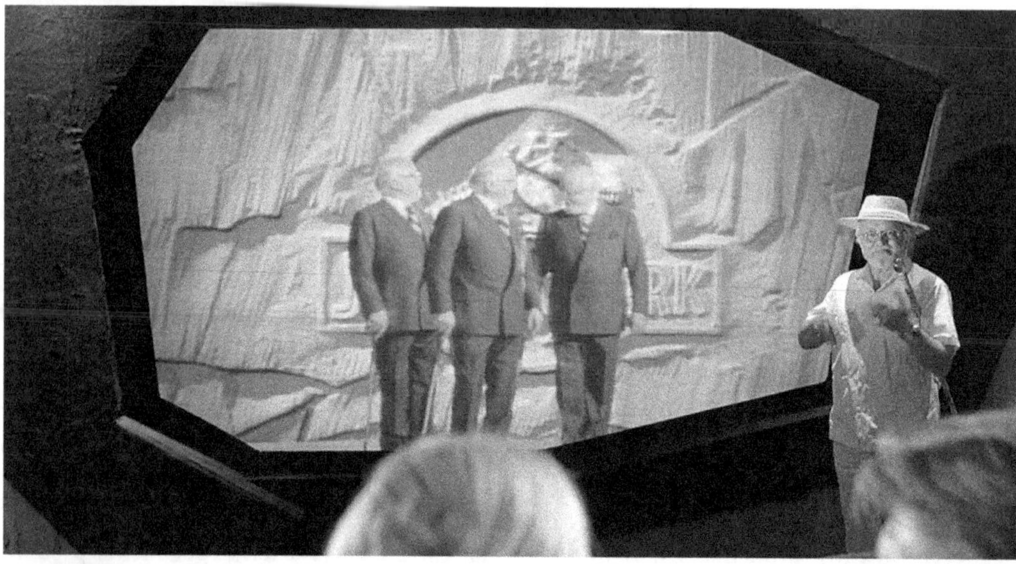

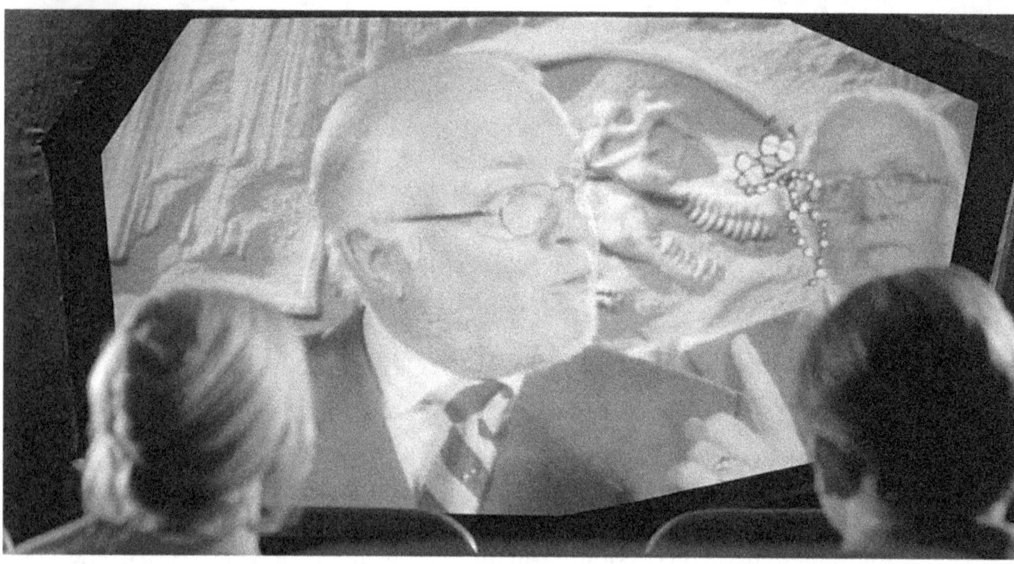

Figures 8 and 9: The cinema of cloning returns to its old tricks.

of that DNA degraded over time, becoming 'full of holes', those scientists used virtual-reality gloves and 'thinking machines' to complete the genomes using spare parts from frogs. Then the cloned embryos are implanted into ostrich eggs to complete germination.

Hammond's experiments attempt to synthesize a smooth multimedia experience for his guests, but, like Kittler's project to recombine film's genetic history, offer instead

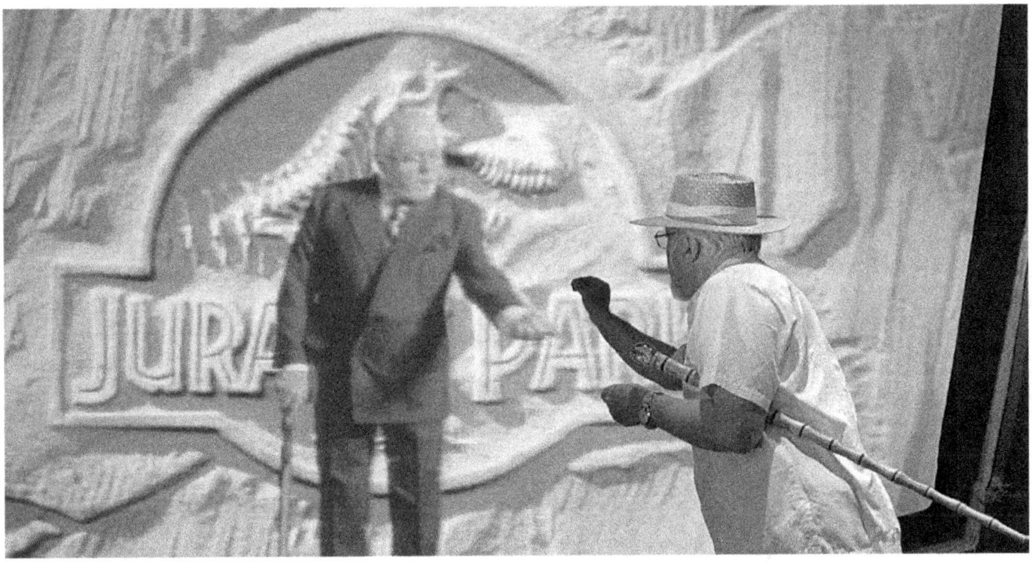

Figure 10: Hammond recreates man.

only a material history of splicing, a series of fragments mutating in a virtual reality. The sequence explicitly reflects on the mediations necessary to create *Jurassic Park* the filmic event. Hammond proudly proclaims that there are no animatronics, commenting on the film's distance from Spielberg's earlier *Jaws* (in which the monster was a robot); the computers used to process and streamline the dino-DNA signal the digital processes that created the film's thrilling dinosaurs. Finally, the matched edits from animation to nature-documentary images of mosquitoes getting trapped in amber construct a seamless history of moving from the real image to the fake image – *but in the wrong order*; here animation precedes straight film, unconsciously suggesting that the authenticity of the film medium lies in its universal special effect of animation. This reversal of chronology unwrites the ordering gesture that allows for Kittler to establish a family tree for cinema and cinema-like images; without an original film for TV to copy or clone, all the media of spliced images appear like an open set of sisters without a patriarch to bind them into historical generations.

Hammond's film takes us towards the origin of embodied life (when he reaches out to take his image's blood, he recreates the imagery of Michelangelo's *Creation of Adam*), but that origin isn't as natural as we would have liked. The origin doesn't even feel particularly momentous: in place of the brutal violence against the Real imagined by Baudrillard, we have here just a tiny pinprick. The punctuation itself comes from the side of the Real, and opens up the image into a series of trick shots – just as the pinprick into a dark room allows the world to recreate itself on the inside of a camera. And the score is lightweight and cartoonish, not the tragic tones that Hammond mentions haven't been composed yet.

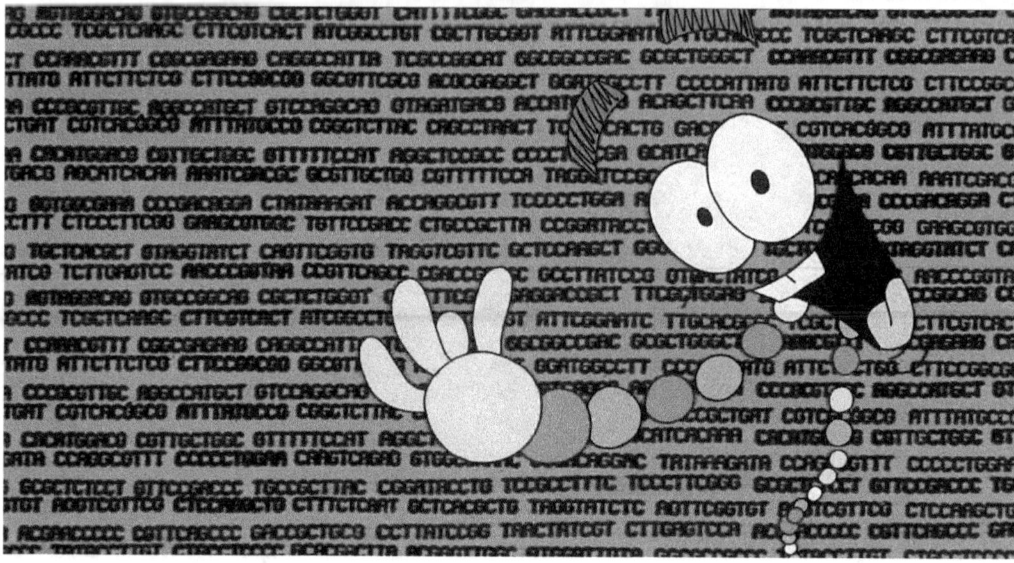

Figure 11: Holes don't damage DNA; they form the basis for its structure.

Everything feels animated, with violence provoking transformation but not death, revealing the fundamental malleability of the cartoon body.

Look again at Mr DNA, who inserts himself between the doubled Hammonds, both as their double and as the medium allowing for their doubling. The quintessential cartoon, his body is made of gaps, manifesting an infinite flexibility – he stretches into the Apatosaurus we've just seen, and stays in constant motion. Even though he associates 'being full of holes' with the effect of time on the dino DNA, Mr DNA himself is nothing but a series of holes – just as 'real' DNA contains countless lacunae and redundancies; far from damaging or denaturing DNA, these holes enable DNA to do its work, to vary itself across countless almost-identical clonings (Figure 11).

Mr DNA presents an image of what Baudrillard struggles with in his work on cloning: the incapacity of 'pure information' to stay still, to remain *the same*, precisely because its purity – its slight materiality – makes it so materially *vulnerable* to data loss and consequent metamorphosis. That's what allows the ancient mosquitoes to hijack dinosaur DNA, and what allows that DNA to be re-cloned in the present, by 'the real miracle workers' – and, as the film will go on to argue, for that DNA to escape the programming of its cloning and reproduce itself. As Jeff Goldblum exclaims, 'life finds a way'. And isn't 'life' just the way that DNA finds in the time-lapse series of partial images produced by each individual life form, their variation within continuity giving time to natural history? What 'finds a way' here seems to be splicing itself – the animating force within film, cartoon and digital images, as if all cinematic media are simply expedient means for splicing to splice off, to mutate, into infinite series of complex remediations.

When the scientists finally view the lab, it appears as the afterimage of its cinematic representation, the movie screen replaced by a glass window. Baudrillard might speak of the 'precession of simulacra' here, the way the 'real' image becomes one in a series of virtual snapshots. But what does it mean to say 'real cinema', here, when amber and fossilization show nature time-sampling itself, and where even evolutionary history (the triumph of DNA's splicing power) looks like a rudimentary film technology? When the scientists break out of the ride and into the lab, it isn't a break at all, but a continuation of the animation begun in the dino DNA and only divided in Hammond's film. Indeed, once the audience arrives in the genetics lab, they've become the film: we see them from the wrong side of the screen. But does it even matter, since the film has demonstrated the permeability of the screen? For this moment, instead of the uncanny tragedy of cloning as the death of the reality principle, the film leaves us with a notion of cloning as the comically formless form of mediation itself – the way one medium, as Marshall McLuhan suggests (1964), always remediates another medium, making cloning and differentiation into the very principle of the material instantiation of information.

Mr DNA urges us to rethink cinema, mediation, cloning and even life itself from the perspective of *animation*. Animation directs our focus to the material structure of the film strip, a series of informational nodes held together by gaps, discontinuities that give them both their consistency and an endless flexibility. From this perspective, Mr DNA obliterates the difference between digital effects and film cinema, since they are both cloning media – cloning media like the DNA strand. Both information and material, DNA conjures continuity and discontinuity, or discontinuity as the guarantor of continuity. Especially for us humans, the clone highlights these genetic lapses that structure our human performance of humanity, as when, at the beginning of the edutainment film, Hammond gets out of sync with his image. 'I've got lines', Dr Hammond exclaims in panic. Leaving aside the fact that he made an informational video that would require him to personally introduce *every single tour*, the force field of clonal media reveals the gaps in Hammond's performance of *being John Hammond* – not so much the way that he doesn't know his lines, but the fact that his performance of being himself is scripted in the first place.

Sianne Ngai discusses a peculiar affect she calls 'animatedness', or the feeling of having any feelings at all – the sign of the capacity to be moved, and to move, that paradoxically distinguishes humans from objects even as it foregrounds the human capacity to be the object of emotional information, giving not only lifelikeness but also threatening agency by demonstrating our capacity to be hacked into by various affective technologies. Ngai suggests that trick film techniques first made this uncanny paradox of animatedness palpable to their audiences (2004: 90); for Scott Bukataman, animation in film explicitly conjures the idea of 'life, autonomy, movement, freedom', even as the techniques necessary to produce it rely on 'the preclusion of the random' (2012: 109). Mr DNA, however, the animated version of our animal being, argues that such 'animatedness' isn't just a trick of film technology, but instead the nature of our DNA-animated being. Our selfhood, what we sophisticates call our 'performativity', is always an animation – a series of images linked by absence, a stop-motion

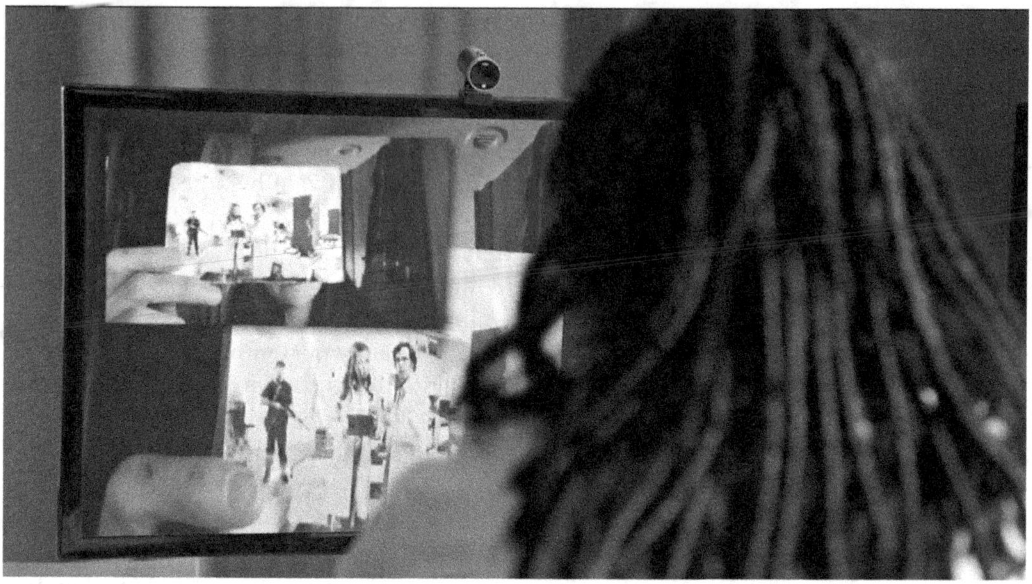

Figure 12: Sarah and Cosima duplicate (and re-duplicate) themselves and their creators.

discontinuity, spackled over an absent continuity of self. The ethnographer Teri Silvio has recently called for us to replace the jargon of 'performance' with 'animation'; in her words, the appeal to 'performance' makes us 'lose sight of the uncanny "illusion of life" that makes [animated] characters appealing, of their particular blend of materiality and imagination, and of their diffuse agency' (Silvio 2010: 423).

Silvio's words almost perfectly capture the 'appeal' of *Orphan Black*'s most spectacular special effect – Maslany's performance, the animation of her body, with its unique ability to diffuse its uniqueness into an endless series. Just as invisible as the ubiquitous trick shots, Maslany's acting, Maslany being Maslany, by which I mean Maslany being not-Maslany, also makes the show visible in the first place – without her unique capacity to lose her individuality in her characters, there would be no show. Framing her, *Orphan Black* obsessively returns to the way in which cloning technology clones already-extant technologies, by surrounding her doubles with endless technologies of image-making. Garrett Stewart observes that, in science-fiction film, 'cinema becomes a synecdoche for the entire technics of an imagined society' (1985: 161), with camera-tech standing in for whatever technology we might happen to imagine. *Orphan Black* shows that this is no accident: film and digital cameras, in cloning the ur-tech of genetic inscription and variation, allow us to visualize the invisible special effect of DNA-animation as it works within actors and the roles they animate. In other words, *Orphan Black* will radicalize *Jurassic Park*'s genealogy of cinema by splicing film into the human body, inscribing its capacity to remediate as an animating force within our natural history.

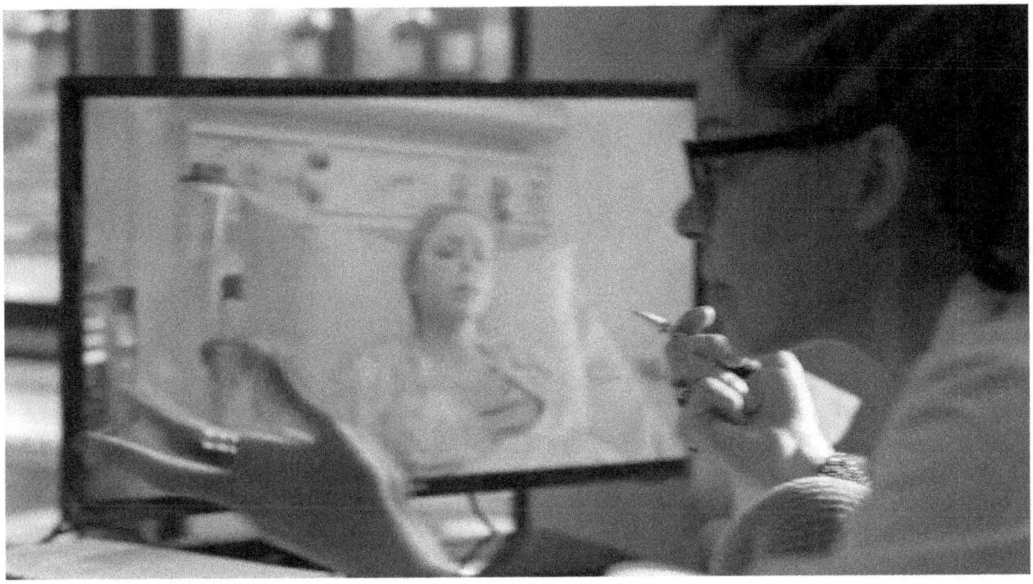

Figure 13: Sarah and Cosima are interrupted by the posthumous image of the woman who lives on in their genetic code.

In a typical sequence from early in *Orphan Black*'s second season, we see Cosima and Sarah discussing their hybridity while hybridizing their reality (Figure 12).The sequence visualizes the two Maslanys' origins, as Sarah shows Cosima a photo of the original Project Leda scientists who created them. Communicating by webcam with Sarah, Cosima explains the mythological significance of the hybrid fertilization of Leda while re-photographing the photo as a screengrab, a kind of hybrid image, to examine later (I, then, re-grabbed Cosima's grab to produce the above image). But Cosima and Sarah aren't alone – Jennifer, another clone who has already died of a mysterious disease, is watching from Cosima's other monitor, and Sarah's daughter Kira waits outside for her mother.

Orphan Black likes to use overlapping screens to visualize its clones, probably because it helps save production resources. But in these virtuoso scenes that film the show's own means of film production, which splice its splicing into visibility as the open secret of its origin, *Orphan Black* argues that – from the vantage point of the genome, as from the vantage point of the cinematic image – all these technologies clone each other. If the webcam can sample the photograph, it's only because the photograph already could sample time; if the clone can sample reproduction, it's only because reproduction already could sample the body. Hybrids at the origin, doubly missing the human, the Leda scientists overcome human nature by rediscovering the animation of animality: making Maslany, originally, goddess, woman, beast, monster, replicative genome, an endless proliferating list of self-sampling images. To that list, we might add that Maslany becomes photograph, film, video,

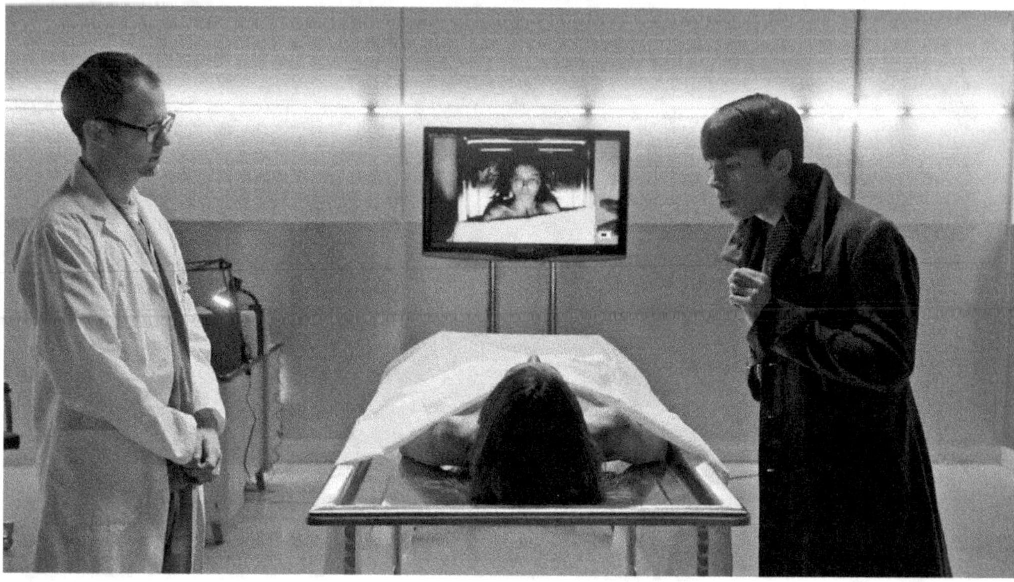

Figure 14: The real-time video monitor splits the mortuary into a space of suspended animation.

television, freeze-frame: in *Orphan Black*'s media history, all these technologies of imaging or projecting the body, with their minor, mutable materiality, become a precession of clones animated from within by the self-subversive force of DNA's many holes.

Before we ever see the clones together in one image, *Orphan Black* shows us the ways their image, their identity, is always diffracted and diffused, as a sustained series of cloned animations. In the pilot episode, Sarah's foster brother Felix goes to a mortuary to misidentify the corpse of the dead Beth as Sarah (Figure 14). Corpses, like photographs, are partial images of a body that can be misidentified, dislocated and re-imagined. *Orphan Black* radicalizes this filmic essence of the corpse by doubling Beth's body with a real-time image on a DV monitor. Even in its stillness, the DV film shows that Beth's image was already on the move, animated by life's capacity to extend and remediate itself in a comic series of endless transformations, endless reanimations. Maslany's body, meanwhile, shows itself as a site of provisional, shifting, overlapping animations; the matrix of mediation and cloning, she dares us to see her body *as* science-fiction, as the animating cloning that dissolves the line between real genetics and special effects.

Watching While (Face) Blind: Clone Layering and Prosopagnosia

Sharrona Pearl

This essay will explore how the identical clones in *Orphan Black* are rendered distinct in ways that do not rely on their facial features. It will discuss whether the world of *Orphan Black* renders all viewers essentially face blind by examining the viewing experiences of those with clinical face blindness, known as prosopagnosia, and by arguing that when it comes to telling the clones apart, they have the advantage, surprisingly, over neurotypical audience members. We'll delve deep into the multiple ways in which this distinction is communicated, but we must acknowledge at the outset the function that acting itself plays: Tatiana Maslany does a superlative job in a very demanding role.

Maslany's portrayal of all the clones as distinct (if identical) characters, with distinct (and not identical) mannerisms, body language, self-presentation choices, accents and speech patterns is nothing short of magisterial. Don't take my word for it, or the words of the endless fawning critical reviews, fan feedback (#cloneclub, #clonesbians), and awards nominations and prestigious wins (including a 2016 Emmy).[1] Consider instead what it is like to watch *Orphan Black* with prosopagnosia, which is the inability to distinguish among faces.[2] For people with this neurological condition, watching many television shows is a highly puzzling and deeply frustrating experience. When all faces look the same, distinguishing characters and following plots is really quite hard, leading to a strong preference for superhero narratives and other fictions with highly identifiable and visually distinct characters. *Orphan Black* ought to be amongst the hardest of them all, as distinguishing each clone is fundamental to the story.

Instead, it's easy. It's easy because the main characters are all identical – and *all* viewers, not just the face blind (for whom the identical faces in this show are essentially the same as the distinct faces in other shows), have to tell them apart. So the characters are very carefully and very explicitly distinguished not by their facial features, but by everything else. In fact, face-blind watchers of *Orphan Black* may well have an advantage over neurotypical viewers: they are already used to telling people apart in ways other than their facial features. Some of them have even become quite good at locating and paying attention to a completely different set of distinguishing clues. In the world of *Orphan Black* viewership, we are all, in a way, (almost) face blind. In a world where (almost) everyone who matters is identical, facial features become (almost) irrelevant. Almost, because the fact of character identicality matters very much indeed. That's the whole point. In *Orphan Black*, while distinguishing identity is very much at stake, the identicality of the characters (almost) levels the playing field in the game of identification: neurotypical and face-blind viewers all have to turn away from facial features.

The stakes are lower, of course, for neurotypical viewers, who can simply visit through fiction a world in which facial features are not the means of identification; as fiction, that world remains just a game. Our assumptions may be challenged, but our frameworks are not overturned. The possibilities of reframing our ways of knowing others, of undoing the primacy of the face in identification, remain unexplored. We do not yet encounter and accommodate others as incomplete, interconnected, unwhole, in-process – which is the promise of a world in which we do not rely on faces (Edkins 2015: 126–27). Such an encounter, as Jenny Edkins shows, would be traumatic, even brutal; it would force an acknowledgement that our foundational belief in the power of the face to communicate unique identity is false. The realization of any utopia necessarily requires the pain of dismantling the systems in which we are utterly complicit and entirely imbricated. The clones of *Orphan Black*, by exposing and insisting on alternative modes of identification beyond the face, lay bare some of the systems of facial hierarchy discussed by French philosophers Deleuze and Guattari (1987: 115). For the face blind, this exposure, even in a world that still insists on identification (if by other means) is easier to navigate than when those systems are obscured.

The accessibility of *Orphan Black* for face-blind viewers reveals just how much we rely on faces to distinguish character(s), and how many other markers are needed to distinguish character when faces are all the same. Non-facial identificatory clues are particularly relevant when one clone is imitating another, in the so-called 'clone swaps'. These moments are exciting, and they reward the careful viewer for picking up on the clues and slippages between (identical) clones. And ultimately, these moments highlight how very reliant we are on facial features unless other aspects of character are very carefully established and emphasized. The face-blind world of *Orphan Black* is not a world that transcends the visual or transcends identity distinctions, nor is it a world that overcomes the primacy of what we see in establishing and communicating identity. In fact, although *Orphan Black* complicates visual identity markers, it ends up underscoring their importance, even as it shifts their location from facial features to other visual and verbal markers. The world of *Orphan Black* is still a world of visual identity systems; it just happens to be one in which face-blind rather than neurotypical viewers can have the upper hand.

Orphan Black is successful for many viewers with prosopagnosia because the embodied presentation of each clone is very different. To someone with face blindness, generally attractive people with similar hairstyles and colours, with basically the same skin tone, and with overall consistent body types (which is to say, in most North American television, young, thin white people) are functionally indistinguishable. The female clones, despite having exactly the same facial features, can be easily told apart. But if that isn't enough, they are quite literally *told* apart – the show's evolving take on the classic gay male best friend, Felix, very carefully narrates and repeats the story of each clone to make sure we understand.

Orphan Black lends experiential insight into our own viewing and identification practices. A world of clones turns us all into face-blind viewers, but bad ones, because neurotypical viewers haven't developed other ways of paying attention to people, which often include highly developed differentiation of voice cues and expressions. *Orphan Black* presents an alternative system of visual identity, one that playfully challenges the primacy of facial features but does not challenge the need for visually differentiating people. This is most obvious in the clone-imitating-clone scenes and the scenes of multiple clone interactions, which force viewers to reflect on how it is we *do* distinguish the clones, and hinting at what it might be like to be in a world where facial features themselves are meaningless. We should consider how, in *Orphan Black*'s world of essentially meaningless facial features, differentiation *is* communicated. Many of the effects are achieved through technology and its deployment, in which make-up and hair and accents and voice cues and carriage act as technologies of identification alongside camera techniques and screen shots. This chapter explores the expansion of technologies of identification when facial features fail. In short, we find other ways to distinguish people. Neurotypical viewers just aren't as good at it as are the face blind, because this is what the face blind have been doing their entire lives.

Will the Real Clone Please Stand Up?

What is Tatiana Maslany's starring role?

Despite playing multiple characters, Maslany is perhaps mostly recognized as the actor who plays Sarah. We've been trained to see Maslany primarily as Sarah both because that's how we first meet her and because Sarah largely drives the plot and the narrative. Viewers deeply immersed in the world of the show often forget that the other clones are also portrayed by actors, or, rather, one specific actor. For the face blind, this is even more extreme; as one viewer writes on a Reddit post about *Orphan Black*:

> [...] to me, these might as well be different actresses. I have to keep telling myself that Helena IS the same actress as Sarah and Rachel, so it's silly to say the Helena is my favourite. Lately, I've gotten to be more of a fan of Allison, too ('I... need to... cut something!'), and have some moments of total brain farting, wondering what happened to the actress that plays her (maybe I should go look her up to see what else she has acted in). Head-smack!
> (Reddit 2016)[3]

We know, of course, that Maslany plays them all; every article about her marvels about it, even fawns over it (and rightly so), but in the end, it is too hard to imagine her as many (different, clone) people when she is only one.[4] There is a kind of invisibility in that; the actor who plays Cosima and Alison and Helena and Krystal and Katja and Tony and others disappears, leaving behind only Maslany-as-Sarah. But maybe that is not surprising; maybe it is an entirely predictable outcome of the fracturing of one face into so many parts. What's more interesting is how much we stop paying attention to facial features in the world of *Orphan Black*. Because beyond telling us that someone is a clone, the faces of the clones (and this is counter-intuitive, different from everything we imagine we know about faces and their communicative power) tell us nothing at all.

It's the other stuff we pay attention to: the hairstyles and the accents and the make-up and the clothes and the gestures and the gait. Within all those things, as the show acknowledges again and again, is the history of the character's class, geography, personal experiences, ideologies, beliefs, relationships to sexuality and gender, as well as their encounters with others. Helena's scarred back tells us something about her; indeed, it tells us more about her than the face she shares with so many others. This is true of Sarah's punky smeared eyeliner, split ends and purple streaks, Beth's posture – which Sarah has to work hard to emulate – and Cosima's cool-kid tattoos. Cosima's glasses tell us that perhaps she eroded her vision with all that scientific, nerdy reading. According to epigenetics, all the clones could have the same genes expressed with different phenotypes, which might be responsible for Cosima's poorer vision, but the symbolic power of glasses on the studious clone works to reinforce the nurture rather than nature assumption that immediately leaps to mind.

All of these factors are part of what Deleuze and Guattari call the 'abstract machine' that contributes to the coding of the face (1987: 141–45). The face itself (its features, its colours), they claim, is meaningless on its own; it is the other stuff we pay attention to that tells us something about the person, which we then inscribe on to the face itself. And of course, it's our relationship to all that other stuff (including history – theirs and ours) that determines how we code the face in question as a deviation from the ideal (white, western, male, typically Christian) face against which it is always measured. (Some histories have been made universal – see Rushing later in this book.)

So, what is an identical clone imitating another clone with a different accent, different gait, different history, different set of assumptions and experiences and relationships to do? Bear in mind (because we so easily forget, and indeed we are *supposed* to forget): Maslany could have one clone be another by simply *being* the other clone, since she is in fact all of them. But the other point is that Maslany is all the clones, but all the clones whose faces are the same are actually entirely different; what the show demonstrates is that having the same face isn't even close to enough to convincingly portray another person. Every clone swap has to acknowledge the identicality of the face while revealing all the differences between clones; it has to do this while performing the imitation well enough to drive the plot forward, but with enough slippages and mistakes to keep the tension high, the moments funny, and the audience feeling expert enough to detect these leaks while the other non-clone characters in the show (ideally) do not.

Deleuze and Guattari's abstract machine is virtually the playbook for these moments when one clone abandons her own distinguishing non-facial traits to take on another's. The clones are good at it (some better than others); women know well how to manipulate their appearance, and women know well how appearance itself is a way to become anonymous, to have personhood evacuated, to have identity become general and a function only of the way one's face is coded and the face is performed. The show itself plays with the idea that the way we look determines who we are, especially if 'we' happen to be women. It doesn't determine who we are in the world of *Orphan Black*; the army of female clones are importantly not interchangeable and not indistinguishable – even when they want to be, their individuality peeks through the cracks. (And as we see when we meet the male clones who in fact are an army, it's different for men: they really are an indistinguishable killing clone machine, partly because the military makes them one, and partly because, as men, they just aren't as good at manipulating how they look and performing their faces. They've never had to.) To paraphrase Cosima's famously meme-ified statement in Episode 2.2 ('Governed by Sound Reason and True Religion') that 'my sexuality is not the most interesting thing about me', we might understand the show itself to be saying about the clones that 'our identical faces are not the most interesting thing about us'. Though they are indeed the conditions of possibility that makes everything else so interesting, including when one clone is pretending to be another.

The fandom and critical term that has emerged for clone-imitating-clone (-imitating clone, in the case of Helena as Sarah as Beth in Season 1) is 'clone swaps', though Maslany's own description of 'double layer' may have more theoretical heft. As Maslany notes, some clones have more aptitude for these imitations than others: Sarah's background as a con artist and minor grifter has prepared her well for the acting and manipulation required to succeed, and Alison's community theatre efforts have pay-offs besides comic relief (Zimmerman 2015). Sarah maybe does it best of all, but not seamlessly – we can still see Sarah beneath whichever *sestra* she's inhabiting. If we couldn't – if the swaps were flawless, if there was in fact no layering at all, if Maslany just acted Sarah-as-Alison (say) as Alison (which she can of course do), then the whole premise of the show would be undermined. We're supposed to forget that each clone is played by the same person. Without the layering, we'd be forced to remember. It's in fact not a swap, not an exchange, and not even a performance – not exactly. The term *layering* is exactly right: we are meant to see the clone-beneath-the clone. We are meant to search for – and find – the edges sticking out, the bumps and the thickness. We are meant to understand that each clone is distinct, and not just in self-presentation but in the entire way of being the self. We are meant to know that none of them can entirely disappear into another person. That is central to the whole: even biological clones are (by nurture) unique, (through history) individual and (from experience) different. That's the theoretical part. The practical part is how Maslany does it, and what it says about gesture, embodiment and the communication of identity.

How It's Done

Layers, again. 'How' is a question of technicalities, of course: camera angles and body doubles and make-up application and wigs and all kinds of cheats and shortcuts. There's a camera ('TechnoDolly') that exactly memorizes and repeats the shot done with Maslany portraying one clone and body double Kathryn Alexandre (who also plays the Prolethian child-minder) as another. Maslany then shoots it again as the other clone, filling in for Alexandre's work with her initial positions as the other clone already recorded[5]: The details of execution matter both for their own sake – they are really interesting, and emerged partly from budgetary rather than artistic reasons, as there wasn't money for motion-controlled camera moves (Edwards 2015) – and because they make us attentive to the myriad people involved along the chain. The technicalities don't happen by themselves. The technicalities require a whole set of expertise and creativity and money (though with rather less money comes more creativity) and technology and skill. The technicalities make celebrities not only of the actors but of those around them: the make-up and hair artists, the body doubles who are actors in and of themselves, the camera operators and editors. The cameras, too, are celebrities ('TechnoDolly'), and also part of the technology.

These methods are not secret: many of the cast and crew from the show have discussed at length the make-up and hair transformations that bring the various characters into being, giving us the tools to explore the cheats and slippages that go into layering one clone upon another and shooting one (or two or three) clone(s) with others. There are acting techniques that contribute to the process, including musical playlists that Maslany has curated for each clone to listen to in order to get in to character. Spotify has made them into easily accessible lists online: Sarah for example gets M.I.A. and The Streets, while Alison the community theatre queen is *Les Mis* and *West Side Story*. Cosima is Grimes (in a shout-out to Canada) and Daft Punk, and Helena is Tom Waits and The Prodigy (Morris 2015). Costuming, hair and make-up, the dialogue and gestures that highlight how one character relates to a given clone – all let us know which clone it is. As one viewer with prosopagnosia writes: 'Mostly I just depended on Sara's [*sic*] reactions to tell me if this was another clone. I couldn't tell Beth was until Sara [*sic*] was freaking out to her friend about it'.[6] And, of course, Maslany's own facility with accents and language helps: her mother speaks to her in German, she went to French immersion schools and she grew up in English-speaking Saskatchewan, Canada (Aurthur 2013). All those pieces are part of the identificatory puzzle. But so many of those particular techniques hinge on differentiation, on creating distinct characters that can be easily visually, verbally and textually distinguished from one another. What about the moments when they are merged? These slippages signal that it is not the face that determines character; that is the world that face-blind people live in all the time.

That's the least of the issues.

The limited initial budget for the series constrained not just camera options but also other effects, including highly expensive prosthetics for Maslany, though other characters did get some additions, particularly Aldous Leekie (Miller 2015). There was also a major time limitation – Maslany had to be able to switch over between clones *fast*, which meant that it really couldn't get too fancy. Make-up artist Stephen Lynch and hair stylist Sandy Sokolowski, themselves now experiencing a kind of celebrity status based on their work for the series, have said they often have only 30–45 minutes to get Maslany into cosmetic character, during which time she's also running scenes with the body doubles; it's hectic at best (Miller 2015). But it still has to be done very, very carefully, from the clone-distinguishing physical touches to the character-differentiating style that the make-up both signals and encapsulates, as Lynch and Sokolowski explain in a joint Yahoo! TV interview (Nemetz 2014). Sarah, for example, the punk, will 'slather on some soot to make her feel different from the others, and to keep her identity'. Alison's make-up is drugstore quality (and bought there) since 'she got her routine down ten years ago and she never varies'. The make-up is a class marker, but it also provides insight into her Alison-ness, as: 'these bright colors that make her feel in control. If you can control your look, to a lot of people, they can keep from flying apart'. While Cosima and Helena's make-up seem to be the most elaborate, it is actually Rachel's face that is most painstaking to apply and is, indeed, the most expensive – exactly like the character herself. Lynch notes that 'Rachel actually takes the longest… I spend the most money on her too, for those high-end cosmetics'. There are other costs beyond the make-up that must be taken into account: Maslany's skin suffers from the repeated chemical application and removal, sometimes turning quite red and becoming raw. Lynch will, at the end of a long day, cheat; rather than taking off previous layers, he tries to do whatever he can to save Maslany's skin and won't always remove what was there before (Miller 2015). In this way, the clones sometimes unintentionally layer, or bleed, into one another, adding a note of complexity to their uniqueness and the way that it is signified through the decisions they make about how they self-fashion. The rigid technical precision required to shoot with TechnoDolly stands in sharp contrast to the unpredictability of the clone layering in both the make-up cheating and Maslany's portrayal; this makes the bleeding all the more complex and nuanced, giving more clues to the viewers, even those that are not intended.

But bleeding, and layering, is like that, even when it is more deliberate. Because characters, even when played by the same actor, take on nuances and lives of their own. As Maslany has noted, she doesn't always know what is going to happen during the double-layering: 'there's a lot more freedom in terms of [that] the mistakes can kind of live and be. The more slip-ups that exist, the more fun we have with it […] There's so much freedom there to make-believe and to play with the levels of it. Alison as Sarah can be broad, or it can be nuanced, or it can be totally off, or it can be really spot on, or whatever. Some of them are better actors than others, so it's fun to play with all those levels and let the characters dictate what happens, and not try to pre-plan it or control it in any way' (Prudom 2015). Maslany does not know exactly what the audience sees, because she herself does not know what is coming; the deliberate clues bleed into the spontaneous ones, making the message more complicated and more realistic. The abstract machine, even when carefully and artificially constructed for fictional purposes, comes to life and becomes messy as the clones interact and immerse themselves into one another through the interaction of actor, dialogue, make-up, hair and other characters. Uniqueness and identity matter, and assert themselves, even in this context where they could easily collapse and disappear. And that too makes it easier to distinguish the clones in ways other than entirely visual. The slippages and the precision combine to bleed and layer – sometimes quite literally, as in the case of hair slipping out of a hat or headband or ponytail in the chaos of the moment.

Hair is an important part of the transformations, and it is not just a matter of switching styles or throwing on a wig. With Rachel's higher hairline, Alison's bangs, Helena's… hair, and Cosima's elaborate dreadlocks, Sokolowski has to get creative. She too sees hair as an index to personality, noting of Alison's ponytail that 'because she's such a wound-up character who always wants to have it together, we can kind of make it so if she's out of control, we have these little distressed pieces coming out'. And, of course, Cosima's dreads aren't a full wig; there's actually a cage inside the pieces that keep the whole structure quite light and easily detached (Nemetz 2014).

Hair can be covered, and often is when one clone is portraying another. Helena-as-Sarah-as-Beth in Season 1 wears a baseball cap; Cosima-as-Alison sports a jaunty (and very un-Cosima) toque. Sarah wears a headband to play Alison, thereby explaining the lack of bangs, and dyes her hair a uniform brown to imitate Beth. Though hair length presents a challenge for Sarah-as-Beth: Paul, Beth's monitor (a figure, usually a lover, assigned to secretly monitor the health and well-being of each identified clone), wonders how 'Beth's' hair had grown so quickly. (This is in the first episode, 'Natural Selection': 'What did you do to your hair?' 'I got it cut'. 'It's longer… something's different'.) (He also wonders where her ear scar went. Scars and tattoos would be easily identifying cheats for viewers on the one hand, but on the other hand, they are very hard to have to redraw continually between takes, though Cosima has two forearm tattoos: a nautilus shell and a dandelion.)[7] These visual cues around the hair – hats, dreads just peeking out and the visible lack of (Alison's signature) bangs masked by a headband – signal to the viewer the layers, even as we know from the dialogue and plot that it is happening. And these slippages between stylistic choices further imbue the hair itself with symbolic power; once you catch a glimpse of Alison's bangs, say, the metonymical function of those bangs becomes even stronger. The bangs are Alison. Alison is her bangs. And the choices leading to, say, bangs, also becomes interchangeable with identity itself. These visual markers represent character aspects of the clone in question; they are not just modes of making distinctions but extensions of identity and choice.

The visual cues make it more fun, and also help us to maintain our immersion in the show; of course Maslany *could* have each clone seamlessly portray another, though that would undermine the premise and fiction of the show's world. Seeing these little fault lines helps us continue to pretend that the imitations are a challenge for the clones, while at the same time reminding us that really, they are not. It's also kind of a game, of course – spot the swap! – and part of the game is predicated on how very subtle these layering clues may be. (Not always, of course; Helena, for one, is not subtle.) We the viewers congratulate ourselves on figuring them out, thereby yet again forgetting that they are inserted rather than natural, yet again forgetting that the layers are there to help us do precisely that. And most of us need the help.

It works. The layering, when used sparingly and smartly, works to remind us about our own work of forgetting, our own immersion into the show. With all its various manifestations, it works for both neurotypical and face-blind viewers. Neurotypical viewers need them all; for face-blind viewers, probably fewer clues would suffice, because this is what they always have to do to tell people apart.

Watching While Face Blind

People with prosopagnosia tend to develop a host of coping mechanisms, including being very, very good at distinguishing people by their voices, by their gestures, by the way they hold themselves.[8] They also tend to surround themselves with people who have distinguishing non-facial features – tattoos, unusually coloured hair, notable piercings. They like ugly people, because ugly people tend to stand out. They also have a lot of social anxiety and often suffer from depression; face blindness can be very debilitating for those who experience it when faced with interactions with others.

And people with face blindness tend towards books rather than television, which makes sense: television, with its host of blandly attractive similar-looking generally white people, can be hard to follow. There are some easier to watch genres, including superhero shows, with their bright costumes and simple good versus evil plotlines (*Jessica Jones* and some recent, more nuanced examples notwithstanding), and some shows that are to be avoided at all costs, particularly those that rest on a character being a turncoat or double agent. It won't make sense if you can't actually identify the face of the traitor as that of the hero.

In *Orphan Black*, the characters do not just look similar but have precisely the same features. For people with face blindness, that's no different than any other show – except that in this world, that's true for all other viewers as well. The show does the distinguishing work for you. Amazingly (there is something for everyone out there), there are sites and forums dedicated not just to prosopagnosia, and not just to consuming media with prosopagnosia, but to the experience of watching *Orphan Black* with prosopagnosia, with specific guides and clues for the parts that may be tricky for the face blind. The guidance parts of the posts are really rather short, because many people with prosopagnosia love watching *Orphan Black*. It's easy for them to follow. As one face-blind viewer notes on Facebook: 'I've started watching Orphan Black (which, spoiler, features clones), and I'm finding it really hard to process that these people are supposed to look identical, when for me they are *easier* to tell apart than most thin pretty white actresses'.[9] Another echoes that 'The same actress plays many characters, but they all have different clothing and hair styles to distinguish them! I tell them all apart so easily! Each time a new one is introduced, I ask my boyfriend "does she look the same?" and his answer is always "yes". To me they're all soooo different looking!'[10] Face blindness can be an advantage in the clone world.

Because of the show's lack of emphasis on facial differentiation as a means of distinguishing character, there are ways in which it is *easier* for the face blind to follow than for those who rely heavily on features to make sense of who is who. In fact, another Facebook commentator claims that 'I don't worry much about who's a clone, because pretty much all female main characters will be. Because they change their hair etc. I can tell them apart better than neurotypicals, and that's great!'[11] *Orphan Black* may well make real what mildly face-blind scholar Jenny Edkins proposes to be the advantages of prosopagnosia, namely another way of accessing the world that has its own benefits and carries within it a different politics. She suggests a way of understanding prosopagnosia not as a lack, but an opportunity to encounter people on different terms (Edkins 2015: 126).[12] And, as the Facebook commentator above shows, in a world that is differently oriented, it is not just the super-recognizers but also the face blind who have superpowers. Another face-blind blogger, Emmeline May (who goes by 'Rockstar Dinosaur Pirate Princess'), writes in depth about the particular skills she has that balance the challenges she faces: 'While the initial revelation [of my diagnosis] made me feel like everyone else had been walking around with a superpower (the power to recognise people) that I didn't have, I am also realising what the benefits are of my brain compensating for my lack of magical recognition ability. I can read facial expressions, micro-expressions, emotional states and body language LIKE A BOSS'. To compensate for her face blindness, May has developed coping mechanisms to distinguish people in ways other than their features. She pays attention to non-facial clues, which helps her make social decisions, though this cannot compensate for the basic ability to tell people apart by their faces. These sometimes-eerie skills offer unique advantages, such as accessing depths denied to those who are used to reading surfaces: 'I often know how people are thinking or feeling before they even express it. (Which, I am told, is also really annoying sometimes.)' There are both personal and global benefits to a form of relationality in which facial features are irrelevant: 'It maybe makes me poor at initial meetings, but it's helped me be a great communicator. It also helps me to be nicer to complete strangers. You never know, that person pissing you off on the tube might actually be the person who interviewed you for a job, so let's maybe NOT call them a shitweasel' (May 2015). There are daily challenges in not being able to look at people and know who they are.

The Facebook poster above who finds the clones easier to distinguish than actresses in other shows emphasizes this point, as 'one of them I find attractive, and several others I don't like at all!'[13] For her, attractiveness is truly divorced from facial features and based on other aspects of appearance, identity and behaviour; some are more attractive to her than others, even though they are identical. The fantasy that attractiveness is based on character is quite real for people with prosopagnosia – in television and in real life. It's a different – and, according to Edkins (2015), a deeply valuable but undervalued – way of relating that this fictional show makes possible for the face blind.

And it does it deliberately and thoughtfully by the way the characters are constructed and portrayed. The visual cues distinguishing the clones are clear. The accents and even the cadences are utterly distinct. The bodily gestures are totally separate. As face-blind blogger and well-known author S. L. Huang writes:

> In fact, since the show asks all viewers to differentiate the clones based on things other than face, it might be even easier to watch than other shows – I have a lot of trouble on some shows with people of the same gender/age/race having similar hairstyles and I mix them all up, whereas here, they have to make sure that doesn't happen or everyone will mix them up.[14]

Because of the emphasis on distinguishing the clones, they become not only distinct from one another, but even more (non-facially) distinct than other characters, as one face-blind viewer notes on Reddit: 'One funny thing about *Orphan Black* for me is that they put so much effort into emphasizing the differentness of the various clone characters that they become the easiest to remember and tell apart' (Reddit 2015). To put it another way, the (main) clones are on the whole more developed – dialogically, physically, in terms of their experiences and backstories and relationships – than most of the other characters. This is notable because they are all played by Maslany, but it is also notable because many of the other characters are men. Rare is the show where men are less developed and have less depth than the women.[15]

The greatest challenge for face-blind viewers is figuring out that there are characters who are meant to be clones in the first place. This particularly matters (as all the forums are careful to explain and clarify) in the opening sequence when Sarah comes face-to-face with Beth; neurotypicals understand much more quickly that Sarah is shocked not just because she is witnessing a suicide, but because it is being done by someone with a face identical to hers. Huang notes that '[o]nce you pick up on who the clones are at the beginning, which the show is very, very clear about pointing out in multiple ways, it's smooth sailing'. It's best to know in advance what the show, and especially that first scene, is about:

> First of all, I'm glad I went in spoiled for knowing the show was about a bunch of clones and that Sarah gets sucked into the world by one killing herself in front of her – I never would have known why she was reacting with surprise to Beth otherwise, and it was helpful at first to assume every new female character that Sarah reacted to with shock had her face.
> (Huang 2014)

Sarah's surprise is important not just for this initial scene; it becomes a relational code that signals and reinforces cloneship encounters throughout the series. Though this first scene is challenging for viewers with prosopagnosia, even here the show is careful to establish and produce a non-facially-based method of helping viewers identify clones. While Sarah may well exhibit shock for a variety of reasons, there is a very particular kind of 'seeing another clone' shock, sustained throughout the show, that is another way to let viewers know that two characters have the same face. But if visual recognition of cloneship eludes viewers and the shock signal isn't strong enough, the dialogue cues it right up in subsequent scenes. Sarah explicitly lays out to Felix, in words, that she saw her exact likeness jump in front of a train. But waiting for the dialogue takes time, and sorting it out takes effort: blogger Rockstar Dinosaur Pirate Princess comments that 'it took me most of the first episode to work out that the whole thing at the beginning was Sarah looks [*sic*] at the woman jumping in front of the train and recognizes her own face. This was meaningless to me. As far as I was concerned they were completely different people' (May 2015). But she did work it out; the script ensures that she will. It's not just Felix (whose role of gay best friend is much complicated and deepened throughout the show) who plays the explication role: all the non-clone characters do the careful and painstaking work of letting viewers know who is who, and recapping what we need to know, so that the visual clues need not stand on their own. It takes a little longer, and some key spoilers help, but even absent those, the plot can be followed. (I never noticed, but I was too busy noticing the faces. Then again, I can notice the faces. Others can't. So they notice other things.)

But in the End ...

There are plenty of other things in the show for people to notice. There are plenty of other ways to signal individuality. We've learned about what many of those are and how they work, and we've thought about why they are necessary and what new possibilities they may raise. But what they don't do, what the show itself doesn't do, is eradicate individuality; in fact, it does precisely the opposite. Faces are meaningless as a way to differentiate characters because there are so many other ways for the clones to be told apart, because they are so very different. So very individual. And can so easily perform the others, with only the subtlest hints peeking out through the layers. Ultimately, both neurotypical and face-blind viewers are relying on similar and overlapping systems of signification, division, classification, organization of people. These are not coded into the face itself, and in this way the logics of the show do approach the Deleuzian fantasy of seeing people not through their faces. But these systems of signification are still highly political, are still deeply imbricated in the 'abstract machine' that produces hierarchy from the stuff of individuality.

The site onto which the politics of seeing are mapped has shifted, certainly, and in ways that highlight the unique skills that people with prosopagnosia have developed. While some neurotypical viewers may well miss at least some of the layering clues that manifest during clone swaps, face-blind viewers almost certainly don't: it is those clues that they rely on all the time. It is those clues that help frame how they relate to others, clues based on history and experience and their manifestation in the body and the self. That's valuable. That may well be a model for an entirely new way of engaging with others. But it isn't, in Deleuze and Guattari's terms, a dismantling of overcoding systems of signification. It's a shift, and one that gets away from the strictest reliance on facial features themselves, while still heavily relying on representational systems, signification and the dividing and classifying of humans. The show teaches us that faces are useful but not essential markers of differentiation.[16] But I would say that the other markers are probably better and more inclusive, presenting great possibilities in terms of how we know – and judge – others.

Notes

1 For a full list of Maslany's awards to date, see www.imdb.com/name/nm1137209/awards.

2 Recent research on prosopagnosia situates it as a spectrum; some people have total face blindness, while others have milder or differing forms. There is also some indication that various therapies can assist in identification for some along the spectrum. For an overview of prosopagnosia, see www.faceblind.org/research/index.html. The other end of the spectrum includes people known as 'super-recognizers'. For a recent discussion of super-recognizers, see Keefe (2016).

3 The use of Internet data is a much-debated topic; for this essay, I follow the guidelines outlined by the Association of Internet Research (AoIR), which emphasize protecting people from possible harm that may be engendered by the reproduction of writing to which they do

not explicitly consent. Some websites have a higher expectation of privacy than others; more public sites, including those dedicated explicitly to public dialogue, can be reproduced and cited directly. Some scholars suggest anonymizing quotes in order to protect the identities of the writers; others offer paraphrasing as another option. The former approach is limited in the protection it can offer, as quotations can be searched and identities easily unearthed. The latter suggestion presents some scholarly challenges in terms of the robustness of conclusions that can be drawn from paraphrased quotes. Many of the quotes in this essay are drawn from Reddit; others come from bloggers writing on their own websites, whose screen names I use only with explicit permission. Both Reddit and the blogs cited here are explicitly oriented towards public readership. However, other material is drawn from a Facebook group, which presents additional ethical complications. The description of the group states that it is 'For people affected by Prosopagnosia (face blindness) who would like to share their experience, raise of [*sic*] the public awareness and contribute to the development of research on the subject'. While the Facebook group is technically public, participants have some expectation of privacy and have not explicitly consented to have their words reproduced in other contexts. For those websites that are more private, including this Facebook group, I have chosen to use direct quotes without offering names or exact URLs. Although the quotations can still be searched, the potential harm to participants is quite limited, while the integrity of their words is maintained. As the Facebook website that I cite has public awareness and research amongst its goals, there is a more outward-facing understanding amongst participants than would exist in a private, closed Facebook group, which I believe justifies the reproduction of the citations verbatim rather than summarizing to make them more unsearchable. For more on the ethical complications of Internet research, see the most recent AoIR ethics report by Markham and Buchanan (2012). For a nice summary of the scholarly debates on the topic, see Wolfinger (2016).

4 See, for example, Lepore (2015).

5 For a video explaining this process, see Gizmodo (2013).

6 Facebook post in a prosopagnosia group.

7 For a video of the make-up team's tattooing process, see BBC America (2013c).

8 For a powerful first person account of living with prosopagnosia, see Oliver Sacks's essay in *The New Yorker* (2010). For a discussion of Deleuze and prosopagnosia, see Edkins (2015).

9 Facebook post in a prosopagnosia group.

10 Facebook post in a prosopagnosia group.

11 Facebook post in a prosopagnosia group.

12 For more on super-recognizers, see Keefe (2016).

13 Facebook post in a prosopagnosia group.

14 SL Huang is currently moving her blog over to another site (Huang 2014). For more of Huang's writings on face blindness, see her website at slhuang.com. She has granted me permission to quote her writing.

15 Felix stands in notable tension with this claim, but as a gay man, he is outside the classic male hero trope. It is worth noting as well that his depth develops over the course of the show; at the outside, he too seems falls into stereotype.

16 With thanks to Sun Ha Hong for this useful formulation.

Part Two

Reproduction/Biopolitics/Community

Game of Clones: *Orphan Black*'s Family Romance

John C. Stout

'I saw a girl kill herself, and she looked exactly like me!' So begins Season 1, Episode 1 of the fantasy about cloning in which sinister motivations progressively deepen over five seasons. At the heart of *Orphan Black* is the Leda clones' quest for self-realization and autonomy; at the same time, the show is preoccupied with the machinations of the various sinister groups who seek to take possession of the clones in order to (mis)use them as part of a grand scheme or experiment. The specific goals of these shadowy organizations keep shifting, as each season of the series features a different antagonist against whom the Leda *sestras* must struggle: the Dyad Institute in Seasons 1 to 3; the Proletheans in Season 2; Project Castor in Season 3; Neolution in Season 4. Finally, in Season 5, as P. T. Westmorland is revealed to be 'the man behind the curtain', the viewer comes to understand that Neolution has been controlling the cloning experiment the whole time. As this 'Game of Clones' has so mesmerized its audience, generating a large fan base, it is clear that *Orphan Black* tapped into the zeitgeist in a unique way.

Why did a show about clones elicit such a powerful response from viewers, becoming an instant classic of cult TV? The clone is, certainly, a potent catalyst for fantasy, as it questions and destabilizes the boundaries between self and other, original and copy. Literature offers numerous celebrated examples of the double and its ability to create uncanny effects.[1] Still, the clone may trump the double as a figure of fascination and dread, since the clone opens up the question of technology's potential to alter human subjectivity. The huge popular appeal of *Orphan Black* can further be understood in gender-specific terms. The principal characters are mainly the female 'Leda' clones, all played by Tatiana Maslany: con artist Sarah Manning, suburban soccer mom Alison Hendrix, neuroscientist Cosima Niehaus, childlike trained killer Helena, power-hungry and psychotic Rachel Duncan, police officer Beth Childs, beautician Krystal Goderitch and paranoid hacker Veera 'M.K.' Suominen, along with several other female clones who do not appear or who make brief appearances in the series' episodes. There is also a widespread perception among viewers and critics that the show's audience skews female. *Orphan Black* appeals strongly to young women in their teens and twenties who find themselves uncertain of their place in the world, one notes in reading commentary about the show online. Such young female viewers feel an intense bond with the struggles of Sarah Manning and her *sestras* to discover who they are, to take control of their bodies and identities and to fight those who seek to experiment on them or even patent them, treating them as less than fully human.

The popularity of the term 'sestras' among the show's main characters and among the fans brings me to the main concern of this paper: the (over)emphasis on family in *Orphan Black*.

Having first encountered the show by watching a marathon presentation of Season 1 on Canada's Space network, I quickly became enthralled, taking great pleasure in the writing, the acting, the art direction, the music – indeed, every aspect of the show's production. Nevertheless, as I continued to watch *Orphan Black*, I became increasingly aware of a heavy emphasis on family relationships in the series. This emphasis on family did make sense, as it both underscored and mitigated the notion of orphanhood. Siobhan Sadler ('Mrs S') is Sarah's foster mother (and, as it turns out, Sarah's niece/sister through Kendall Malone); Felix, her foster brother, often acts as her sidekick; Alison is a 'soccer mom'; Dr Aldous Leekie is a kind of surrogate father to Rachel Duncan, until her real father resurfaces; Susan Duncan, Rachel's mother, makes a surprise reappearance at the end of Season 3, after we have seen Susan's younger self in the Duncans' home movies from the early 1990s several times, offering the image of a blissfully happy young family; Helena is so touched by Donnie Hendrix's making a 'Dad decision' to support his family as a drug dealer in spite of the risks involved that she teams up with him in Episode 3.9, using her formidable assassin's skills to defeat the drug dealers with whom the Hendrixes have, unfortunately, become entangled. ('You should not threaten babies', exclaims Helena in a face-off with those dealers.) I could see that the show was satirizing the traditional family (via the Hendrixes, in particular) while also exploring a darker side to family dynamics, in part through the recurring figure of the 'bad mother'. In Season 3, for example, Bonnie Johansson tells her daughter Gracie, after Gracie has lost the child she was carrying: 'Gracie, that child was the only reason you were allowed back into this family. You will not be welcome in it again!' When the creators of the series, John Fawcett and Graeme Manson, invented the character of Dr Virginia Coady (the tough but misguided den mother to the Castor clones), they referred to her as 'Bad Mother' (Lutes 2015). Rachel's vicious knife attack against Susan Duncan in Episode 4.10 stems in part from Susan's failure to provide Rachel with the mothering she longed for while growing up. Kendall Malone, whose two cell lines were used to create the Leda and Castor clones, initially appears to be another incarnation of the 'bad mother', as she was cloned in prison without her knowledge or consent when she was serving time for having killed her son in law, John Sadler (Siobhan Sadler's husband), with a pair of garden shears. Kendall then becomes a more complex figure in the course of Season 4, as she is transformed into a sacrificial figure when Evie Cho and her minions kill her.

Consequently, we have many sisters (274 Leda clones, according to the list Rachel gives to Felix in Episode 5.10) and several brothers (at least seven Castor clones); a few fathers, good and bad; and a number of 'bad mothers', including even Sarah Manning herself, given that she abandoned her daughter Kira for eight years. While Sarah ran off to engage in dodgy adventures as a con artist, the steadfast, tough and tender Mrs S took on the role of 'good mother' to Kira. Rachel, having failed to take Kira from Sarah, except temporarily, in Season 2, becomes Charlotte's 'new mommy' for a time in Season 4.

Why is *Orphan Black* so obsessed with family? Sigmund Freud's notion of the 'family romance' seems to me a helpful starting point for thinking about this problem. Freud's short 1908 essay 'Family romances' begins with this statement: 'The liberation of the individual, as he grows up, from the authority of his parents is one of the most necessary though one of the most painful results brought about by the course of his development' (Freud [1908] 1959: 237). The child subject's fantasies related to what Freud calls the family romance develop, he explains, at the point when he/she begins to 'doubt the incomparable and unique quality which [s]he had attributed to them' ([1908] 1959: 237). Reaching the understanding that the parents are less than perfect beings leads the child to feel slighted and to imagine that he/she is adopted, an orphan. The child may imagine his/her real parents 'are of higher social standing' says Freud, adding that '[t]his stage is reached at a time at which the child is still in ignorance of the sexual determinants of procreation'. A second stage of the family romance later follows, after the child has realized that '"pater semper incertus est," [the father is always in doubt] whereas the mother is "certissima" [most certain]'. Freud stresses that 'the motive of revenge and retaliation [that is, replacing the imperfect parents with idealized ones] which was in the foreground at the earlier stage is also to be found at the later one'. He underscores the complexities of plot and fantasy that this genre of the family romance affords: 'for its many-sidedness and its great range of applicability enable it to meet every sort of requirement' ([1908] 1959: 239–40). Particularly relevant for *Orphan Black* is the remark by translator James Strachey in his prefatory note to Freud's essay that 'at first [Freud] attributed [family romances] specifically to paranoiacs' (Freud [1908] 1959: 236). A generalized atmosphere of paranoia does, indeed, ground the unfolding of the events in *Orphan Black*, often producing a sense of unease akin to that of *The X-Files* (1993–2002; 2016–18), in which Agents Mulder and Scully slowly uncover a tangled, multilayered government conspiracy. Their investigation of the conspiracy places them in constant danger, as even people who appear to be allies may turn out to be foes. ('Trust no one, Mr. Mulder'.)

The fantasy of being an orphan, viewed from the perspective of the family romance, evokes and evades anxieties about sexual reproduction, identity and origins. The fantasy orphan longs to know who made her and where she comes from, but she also evades such knowledge by attributing to herself an *other* origin, not the conventional one. This fantasy may further conceal/reveal an Oedipal subtext, as it can include the killing off of the real parents so that they can be replaced by idealized substitutes. In *Orphan Black*, we have seen four fathers killed: Dr Leekie is accidentally shot by Donnie Hendrix; professor Ethan Duncan drinks poisoned tea in order to avoid surrendering the original Leda genome to Rachel; Henrik Johansson is killed by Helena and P. T. Westmorland is done in by Sarah after he attacks her in the final episode. Sarah, Rachel, Mrs S and Alison all have conflictual relationships with their mothers. Rachel's attempt to kill Susan is the most extreme case of hostility to the mother here: seconds before being stabbed by Rachel, Susan has just said, 'for all the joy your sisters have given me, for every Sarah, every Cosima, I regret making *you*!'

Moving forward from Freud, I want to consider family dynamics in five more contemporary fictions created by four determining fathers and one mother of *Orphan Black*. Two of these fictions are explicitly referenced on the show, as part of a deliberate intertextual game; the other three fictions belong to a more implicit network of intertexts behind *Orphan Black*. The fictions to which I am referring are H. G. Wells's *The Island of*

Doctor Moreau (1896); Aldous Huxley's *Brave New World* (1932); Ira Levin's *The Boys from Brazil* (1976); Kate Wilhelm's *Where Late the Sweet Birds Sang* (1976); and the Canadian horror film *Ginger Snaps* (2000), directed by John Fawcett. The family resemblances between these fictions and *Orphan Black* are, as we shall see, both enlightening and troubling.

Thomas Henry Huxley's 1894 essay *Evolution and Ethics* provides the best point of entry for a discussion of H. G. Wells's *The Island of Doctor Moreau*. Wells had studied with Huxley and was strongly influenced by his ideas. A popularizer of Darwin's views on evolution, Huxley was nonetheless a more pessimistic thinker than Darwin. In *Evolution and Ethics,* he explores the tension between the pull of nature and a concern for ethics. He argues that 'in the living world, one of the most characteristic features of this cosmic process [by which living beings evolve] is the struggle for existence, the competition of each with all' ([1894] 1989: 62; 4).[2] The 'cosmic process' and the 'state of nature' that Huxley evokes are purely impersonal forces, not guided by ethical motivations. Humanity can exercise some control over these forces but 'the limits within which this mastery of man over nature can be maintained are narrow' (72; 14). Huxley's fundamental point in his essay is to insist that 'the ethical process is in opposition to the principle of the cosmic process, and tends to the suppression of the qualities best fitted for success in that struggle' (89; 31). Huxley also discusses the relationship between humans and animals, noting both similarities and differences between the two; this segment of his essay bears special import for *The Island of Doctor Moreau*:

> For his successful progress, throughout the savage state, man has been largely indebted to those qualities which he shares with the ape and the tiger; his exceptional physical organization; his cunning, his sociability, his curiosity, and his imitativeness; his ruthless and ferocious destructiveness when his anger is roused by opposition.
>
> ([1894] 1989: 109–10; 51–52)

Having observed these links between animal qualities and human characteristics, Huxley nonetheless stresses that 'there is a general consensus that the ape and tiger methods of the struggle for existence are not reconcilable with sound ethical principles' ([1894] 1989: 110; 52). The 'cosmic process' and ethics seem to him incompatible: 'the ethical process of society depends, not on imitating the cosmic process, still less in running away from it, but in combating it' (141; 83). In his commentary on Huxley's essay, James Paradis states that '[in] *Evolution and Ethics*, the Victorian dream of an "ideal man" was thus discarded' (Huxley [1894] 1989: 53); Huxley rejected, in other words, the views of the eugenicists that were becoming influential in Victorian culture (and are reincarnated in the figure of P. T. Westmorland in *Orphan Black*). Paradis notes that 'Huxley concluded that the social interventions of the eugenicists [...] failed to imagine the brutal consequences for the real individual. The selection of humans by other humans, he argued, would destroy the bonds of sympathy that held society together, without which "there is no conscience, nor any restraint on the conduct of man"' (48). Huxley's argument against eugenics would apply equally well to that latter-day version of eugenics called 'Neolution' whose schemes and methods we see vividly illustrated in *Orphan Black*.

The Island of Doctor Moreau is one of the great scientific romances that have earned H. G. Wells the title of 'the father' or even 'the Shakespeare of science fiction' (McConnell 1981: 3). At the end of Season 2 of *Orphan Black*, professor Ethan Duncan gives a copy of Wells's book to Kira 'for a present'. When Kira shows the book to Cosima, Cosima immediately notices that it contains a long, coded message added to it by Duncan, which may conceal the original genome from the cloning experiment that produced Castor and Leda. In Season 3, the book then becomes a major plot device and, finally, a bargaining chip allowing Rachel Duncan to manipulate all the other characters because she alone can decipher the coded message in the book. She then uses it to secure her escape from the Dyad Institute.

What kind of model does *The Island of Doctor Moreau* provide that could help one better understand what is going on in *Orphan Black*? Wells's novel is narrated by Edward Prendick, a 'gentleman' who is rescued by a passing schooner after the ship he is sailing on is struck by 'a derelict' (Wells 2005: 7). On the schooner, Prendick sees a beast-like man whose face 'was a singularly deformed one. The facial part projected, forming something dimly suggestive of a muzzle, and the huge half-open mouth showed as big white teeth as I had ever seen in a human mouth. [...] He spoke slowly, with a queer hoarse quality in his voice' (19–20). Next, on the way to a mysterious island with Montgomery, a man from the schooner, Prendick tells him that he 'had spent some years at the Royal College of Science and done some research in biology under Huxley' (44). Montgomery warns him that '[y]ou'll find this island an infernally rum place' (45). On the island, Prendick discovers more and more signs that something is disturbingly wrong. He is shocked to see that Montgomery's attendant M'Ling 'had pointed ears, covered with a fine brown fur!' (57). Some kind of harsh experiment is being conducted on a puma from the schooner. Then, in the forest Prendick is surprised to see a creature: 'It bowed its head to the water and began to drink. Then I saw it was a man, going on all fours like a beast!' (61). He also sees three human/pig hybrids in the forest, being chased by a 'Thing' that he fells with a stone. (*Orphan Black* explicitly echoes this scene from Wells's novel when a beastlike human creature attacks Sarah by her campfire in Episode 5.1 and she has to drive it off by striking it in the head with a stone.) Prendick reports these encounters to Montgomery, who comments '[y]ou've been meeting some of our curiosities, eh?' (75). Hearing the puma's cries again later on, Prendick exclaims: 'It was no brute this time. It was a human being in torment' (78). Bursting into Doctor Moreau's laboratory, he realizes that Moreau is practicing experiments in vivisection. Later, when a pink sloth creature leads Prendick to another location on the island, he witnesses the Beastmen reciting a series of Laws. The Beastmen seem to have deified Moreau:

> 'Not to go on all Fours; *that* is the Law.
>   Are we not Men?'
> 'Not to suck up Drink; *that* is the Law.
>   Are we not Men?' (91)

'*His* is the house of Pain.'
'*His* is the Hand that makes.'
'*His* is the Hand that wounds.'
'*His* is the Hand that heals.' (2005: 92)

Prendick guesses 'that Moreau, after animalizing these men, had infected their brains with a kind of deification of himself!' (Wells 2005: 92). Moreau finally explains to him that his experiments aim to transform animals into humans: 'Hi non sunt homines, sunt animalia' ('These are not men; they are animals') (104). 'Monsters manufactured!' Prendick shouts (109). The experiments are ethically questionable because of the intense levels of pain that they inflict on the animals. 'Where is your justification for inflicting all this pain?' Prendick asks Moreau (112). Moreau dismisses the question: 'This pain – […] A mind truly open to what science has to teach must see that it is a little thing' (113). He remains untroubled by ethical objections to his work. (In this, he is a powerful model for Dr Leekie, Dr Nealon, Dr Coady and Evie Cho – and of course P. T. Westmorland, to whom I will return.) 'To this day', Moreau tells Prendick, 'I have never troubled about the ethics of the matter. The study of Nature makes a man at last as remorseless as Nature' (115). Considering Moreau's creations at length, Prendick concludes that 'their mock-human existence began in agony, was one long internal struggle, one long dread of Moreau' (150). In the end, the puma escapes and kills Moreau. Montgomery is also killed and the beastmen slowly revert to their original animal state. When Prendick escapes the island and is returned to Britain, he is no longer able to view the people around him as he did before his experience on the island: 'I look about me at my fellow-men. And I go in fear. […] I feel as though the animal was surging up through them; that presently the degradation of the Islanders will be played over again on a larger scale' (2005: 204). *Doctor Moreau* is a work of horror as much as it is science-fiction. The frightening mix of human and animal in the Beastmen is paralleled by the excesses of Neolution and the hideously deformed BrightBorn baby whose birth Cosima witnesses in Episode 4.5 of *Orphan Black*.

The Island of Doctor Moreau can easily be interpreted as a cautionary tale underscoring the dangers and hubris of tampering with the natural order of things and the folly of playing God. It carries ideas from Darwin's *On the Origin of Species* (1859) to their extreme conclusion. In a study of Wells's science-fiction, Frank McConnell explores Darwin's influence on Wells: 'Darwin described an organic world of constant blind flux, struggle, and variation without purpose. What we regard as established species of this process are simply the lucky variant that happen to adapt more successfully to their environment and therefore propagate themselves more efficiently than their competitors' (1981: 55). 'The struggle for existence' (the title of Chapter 3 of *The Origin of Species*) is a major preoccupation of Darwin's work, as it is of *Orphan Black*, and of T. H. Huxley's *Evolution and Ethics*. McConnell remarks that '[f]or Huxley, the lesson of evolution is a bitter and possibly a tragic one. Morality, decency, love – all the things we value most intensely as characteristics of our civilization and our distance from the beasts – can be seen, in the cosmic scale, as mere evolutionary excrescences'

(1981: 60–61). (It should be pointed out, too, in this discussion of *The Island of Doctor Moreau*, that the narrative model which Wells selected for Prendick's story is comparable to that of *Orphan Black*: Prendick, like Sarah Manning after she steals Beth Childs's identity in Season 1, is thrust into a strange and baffling situation in which something is wrong and which he can only gradually come to understand.)

As we learn during Season 5, the initial experiment carried out on *Orphan Black*'s 'Island of Doctor Moreau' (as Susan Duncan refers to this place in Season 4) actually reverses the nature of the experiments conducted by Wells's Doctor Moreau. In Episode 5.5, P. T. Westmorland reveals that he has been conducting experiments on Yanis, a boy whom he had found in a Latvian orphanage many years before. Yanis possesses a unique gene mutation that can produce a miraculous self-healing of the body. This gene mutation has been passed on to Kira and to Helena's babies via the Leda clones. Though physically still human, Yanis has been transformed into a quasi-animal. As he is revealed as the original test subject in the chain, preceding Castor and Leda, he is, then, the first 'orphan black'. He comes, literally, from an orphanage and his existence has been kept hidden, in the dark, until Season 5. On numerous levels, *Orphan Black* both repeats and reverses the main themes of H. G. Wells's famous novel.

Season 1 of *Orphan Black* begins at Huxley Station (which is really Toronto's Union Station) as Beth Childs steps in front of a Government of Ontario commuter train (or GO Train) with the train's colours digitally altered from green and white to dark blue and white. The name 'Huxley Station' obviously recalls the evolutionary motifs in play in *The Island of Doctor Moreau*'s indebtedness to the theories of T. H. Huxley while also paying homage to Aldous Huxley's 1932 novel *Brave New World*, which is foregrounded further in the name of the sinister Dyad scientist and Neolutionist Dr Aldous Leekie. What do the Brave New World of Aldous Huxley's novel and the Brave New World of *Orphan Black* have in common?

Huxley's dystopian novel begins with a quote from Nicolas Berdiaeff on utopias, questioning their value. From the outset, *Brave New World* describes a chillingly impersonal, soulless world in which natural human birth has been replaced by assembly-line 'decanting'; the family no longer exists; a vapid sexual freedom reigns, as 'everyone belongs to everyone else'; social castes are rigidly imposed, from birth; and the wide-spread administering of the drug soma to the population keeps everyone feeling happy.

As the novel opens, a group of students is led into '[a] squat grey building'. They see this: 'Over the main entrance the words CENTRAL LONDON HATCHERY AND CONDITIONING CENTRE, and, in a shield, the World State's motto, COMMUNITY, IDENTITY, STABILITY' (Huxley 1932: 1). Inside the building, '[w]intriness responded to wintriness. The overalls of the workers were white, their hands gloved with a pale corpse-coloured rubber. The light was frozen, dead, a ghost' (1). The 'Director of Hatcheries' leads the students into the 'Fertilizing Room': '"These", he waved his hand, "are the incubators". [...] [H]e showed them racks upon racks of numbered test-tubes. "The week's supply of ova, kept", he explained, "at blood heat"' (3). The reader is quickly introduced to 'Bokanovsky's Process':

> One egg, one embryo, one adult – normality. But a bokanovskified egg will bud, will proliferate, will divide. From eight to ninety-six buds, and every bud will grow into a perfectly formed embryo, and every embryo into a full-sized adult. Making ninety-six human beings grow where only one grew before. Progress.
>
> (4)

'Bokanovsky's Process', we learn, 'is one of the major instruments of social stability' (6). With his vision of this nightmarish dystopian society, Huxley imagines what would result if an extreme scientific rationality were to prevail, combined with a post-Ford capitalist cult of hyperproductivity. The Brave New World leaders also rely on thought control to keep the population obedient, as, via hypnopaedia, propaganda slogans are repeated over and over as people sleep.

The family – and the mother's body, in particular – is absent from this society of the future. As June Deery notes in her article 'Technology and gender in Aldous Huxley's alternative (?) worlds', in *Brave New World*, 'complete ectogenesis, not just in vitro fertilization, is the norm, which means that the site of reproduction is no longer the female body' (2013: 135). Thus, '[t]he biological mother is displaced and her [...] ability to create new life is safely curtailed' (135). In Huxley's dystopian society, male physicians 'have entirely appropriated the maternal function. [...] In fact, motherhood is made taboo' (136). In this novel, 'mother', 'father' or any words associated with the conventional family are considered swear words, words to be avoided, words that make everyone uncomfortable. The state fully controls reproduction. 70 per cent of females are sterilized 'freemartins'. In *Brave New World* and even more so in Huxley's later dystopian novel *Ape and Essence*, 'motherhood is degrading, a curse', writes Deery. This fierce stigmatization and repudiation of motherhood may well be one of the factors influencing the trend in *Orphan Black* towards multiplying the number of 'bad mothers' who appear.

Yet, *Orphan Black* moves away from *Brave New World* even as it echoes its thematics. An affirmation of the importance of the mother–daughter bond is at the heart of *Orphan Black*, in Mrs S's commitment to Sarah and Kira and to the other Leda clones, and in Sarah's bond with her daughter, her foster mother and her *sestras*. Watching the series with the image of Huxley's scary and dehumanized new world in mind, I feel more sympathy for, and acceptance of, the emphasis on family in the show, despite my initial discomfort with this emphasis. Set against the cold and impersonal Brave New World of Mustapha Mond and his fellow World Controllers, *Orphan Black*'s valorization of the notion of family becomes understandable and appealing. Contra-*Brave New World*, *Orphan Black* maintains human individuality and diversity as real possibilities. (The one character in *Brave New World* who truly attempts to resist Mustapha Mond's system and assert his individuality is John the Savage; this character hangs himself in despair at the end of the novel.) As has often been pointed out, *Orphan Black* constantly underscores the dissimilarities among the *sestras*. Each clone is quite different from all the others. The right to be different is precisely what the world of Huxley's novel forbids. Even characters who are only mildly rebellious like Bernard Marx are shipped off to Iceland to be with '[a] ll the people who aren't satisfied with orthodoxy' (Huxley 1932: 252). Sarah Manning's going to Iceland with Mrs S and Kendall Malone in Episode 3.10 perhaps recalls this intertext.

Though they play a less central role in *Orphan Black* than the Leda clones, the Castor clones provide an important link between the show and two famous novels about cloning, both from the 1970s: Ira Levin's *The Boys from Brazil* and Kate Wilhelm's *Where Late the Sweet Birds Sang*. The introduction of the new Castor clone, Ira, in Season 4 implicitly pays homage to Ira Levin and his sinister boy clones from Brazil (as well as inverting the letters of the first name of actor Ari Millen, who plays the Castor clones). As Levin's novel begins, we learn of a plot to assassinate a number of 65-year-old men living in several countries: 'These men they're going to kill, there are *ninety-four of them*, and they're all *sixty-five-year-old civil servants*' (Levin 1976: 27, original emphasis). The plot turns out to be the brainchild of the Nazi scientist Mengele, who has successfully created 94 different clones of Adolf Hitler. Each of these boy clones is being raised by mothers and fathers with the same age and professions as Hitler's own parents. These new parents have no idea who their boys really are. As in *Orphan Black*, a secret cloning experiment is under way but in Levin's novel – the experiment is more deeply sinister.

Another character, Liebermann, is investigating the assassinations. He gradually comes to understand what is actually happening, as he encounters boys who are physically identical in different countries, being raised by different sets of parents. Liebermann consults a professor of biology, Nürnberger:

> 'Mononuclear reproduction,' he told Liebermann. 'Dr Mengele was apparently a good ten years ahead of the field. [...] Mononuclear reproduction. The breeding of genetically identical copies of an individual organism. [...] Haldane, the English biologist. He called it *cloning*, from a Greek word meaning "a cutting", as from a plant.'
>
> (Levin 1976: 140–41)

Levin deploys savage irony as, in the final scenes of the novel, one of the boy Hitler clones orders his Dobermans to kill Mengele. (In a sly bit of dialogue from Episode 4.5 of *Orphan Black*, when Sarah first meets Ira, she says: 'Oh Christ, a Castor!' Ira then reassures her that he is 'a different breed of Castor', to which Sarah replies: 'What? You mean, like a Doberman?') The novel ends ambiguously, as Liebermann destroys the list of 94 names of the boys cloned from Hitler following Mengele's death, even as somewhere, one of the boys plays at imagining himself as Hitler:

> He could hear the people cheering, roaring, a beautiful growing love-thunder that built and built, and then pounded, pounded, pounded, pounded.
>
> Sort of like in those old Hitler movies.
>
> (Levin 1976: 213)

This scene could be read as menacing or parodic. The boy clone is an ambiguous sign in Levin's text: maybe dangerous, maybe innocuous.

The Castor clone Mark Rollins, whom we first encounter among the Proletheans in Season 2, bears an indirect family resemblance to a character named Mark who is at the centre of Part III of Kate Wilhelm's *Where Late the Sweet Birds Sang* (1976), a novel divided into three parts. In Part I, a global environmental crisis is taking place. In response to the crisis, the Sumner family builds a research hospital in the valley where they live. They begin developing cloning research as a means of survival: cloning livestock for food and cloning humans as human fertility rates decline. Echoes of Huxley's *Brave New World* appear, but the view of cloning in Wilhelm's novel provides an absolute contrast with *Orphan Black*'s representation of cloning. While *Orphan Black* continuously insists on the distinctness of each of its clones, *Where Late the Sweet Birds Sang* features groupings of identical clones – 'the Louisa sisters', 'the Nora sisters', 'the Barry brothers', 'the Miriam sisters', 'a group of Ralph brothers' (78–79) – who lack any individuality. Although Wilhelm's emphasis on clones as 'brothers' and 'sisters' prefigures *Orphan Black*'s preoccupation with clone siblings, their group identity becomes these clones' main reality, and individuality is considered a threatening aberration. Part III of the novel focuses on Mark, who, like Mark Rollins among the Proletheans, is an outsider. Both Marks (Wilhelm's and *Orphan Black*'s) incarnate a potentially subversive difference, although Mark Rollins works to conceal his difference until he elopes with Gracie. Wilhelm's Mark openly and defiantly asserts his right to deviate from the group identity, which the Sumner family compound ceaselessly enforces. Again, the contrast with *Orphan Black* here is striking. Where Sarah Manning and her *sestras* create families based on affection and respect for differences, the Sumner family's world comes to resemble Aldous Huxley's dystopian world of imposed orthodoxy and sameness. Near the end of Wilhelm's novel, when a fire nearly burns down the family's mill, Mark is blamed. (This prefigures the Season 2 episode in which Helena – the outsider to the Proletheans' world – burns down the Prolethean farm). Mark finally escapes the valley, taking some of the women from the breeders' compound with him. In the epilogue, twenty years later, Mark and his people return to the valley to take possession of it, now that the family of clones no longer exists. Significantly, Mark's fertile wife is named Linda – a reminder of the Linda character who is John the Savage's mother in *Brave New World.* In the last sentence of Wilhelm's novel, we are told that 'all the children were different' – that is, not clones. I am not arguing here that *Orphan Black*'s creators are consciously differentiating their show from Kate Wilhelm's novel. However, it is striking that, despite their shared emphasis on family, this novel and this television series adopt such opposing approaches to the clone. *Where Late the Sweet Birds Sang* follows *Brave New World* in making the clone a figure of anonymous sameness and repetition, but *Orphan Black* challenges its audience to imagine a different set of possibilities, including possibilities for resistance, as discussed in the other essays in this section.

Before co-creating *Orphan Black* in 2013, John Fawcett had already directed films and episodes of various television shows. The bonds among sisters, which so powerfully anchor the thematics and story-telling in *Orphan Black*, acquire a further layer of complexity if one explores the links between the *sestras'* tale and Fawcett's best-known film, *Ginger Snaps*. Like *Orphan Black*, *Ginger Snaps* foregrounds sisterhood, marginality, secrets, non-normative identities and the disruption of the conventional middle-class nuclear family. This teen werewolf film, a cult classic, reflects Fawcett's enthusiasm for horror – in particular, for the films of David Cronenberg, with their quirky and grotesque esthetic. In his book-length study of *Ginger Snaps*, Ernest Mathijs recounts its genesis:

> The initiative for the film came from John Fawcett, admirer of monster movies and fan of the films of David Cronenberg, who found in the New Zealand film *Heavenly Creatures* (Jackson, 1994) inspiration to start thinking about a story that would explore the relationship between two female protagonists in the face of horrific events.
>
> (2013: 9)

In the film, the Fitzgerald sisters (Brigitte and Ginger) live in a mindlessly dull and narrow suburban community named Bailey Downs. (Not coincidentally, this is also the name of the suburban enclave in which Alison Hendrix and her family live in *Orphan Black*; the Fitzgerald sisters' mother, Pam, bears a certain resemblance to Alison, as she might become in fifteen years or so.)[3] The bond between the two sisters is the main focus of the film, anticipating the importance of the bonds among the Clone Club *sestras* in *Orphan Black*. Like the clone *sestras*, Brigitte and Ginger are marginal figures to their high school peers. 'To the rest of the school', Mathijs writes, 'the Fitzgeralds, with their weird, Goth-ish clothes and morbid tastes, are freaks' (5). The ties between the sisters are tested, and become strained, as Ginger begins slowly to transform into a wolf after being bitten by a wolf-like beast. In the end, Ginger dies and Brigitte is infected with the lycanthropic virus. (The auto-immune disease affecting several of the Leda clones may echo the preoccupation in Fawcett's film with infection, mortality and the struggle to survive.)

Mathijs, in his discussion of the film, constantly emphasizes

> the togetherness of the various fan communities that rally around *Ginger Snaps*. The 'together forever' slogan [that defines the 'pact' between Ginger and Brigitte] and notions of 'bonding', 'pact', and 'pack' as well as mentions of 'sisterhood', or 'family', compose [...] almost an incantation that connects affective communities.
>
> (2013: 105)

In this regard, the response to *Ginger Snaps* richly prefigures the response from fans to *Orphan Black*.

A further twist in the relationship between the film and *Orphan Black* appears in the sequel to *Ginger Snaps*, *Unleashed* (2004) (for which Fawcett acted as executive producer rather than director). In *Unleashed*, Brigitte, 'still infected with the werewolf virus', lives in motels and temporary accommodation, taking monkshood to control her transformation into a wolf while Ginger 'still whispers to Brigitte in dreams, visions, and hallucination' (Mathijs 2013: 118). For *Orphan Black* fans, *Unleashed* has particular relevance:

> The personification of *Unleashed*'s destruction [of the core ideas of *Ginger Snaps*], and of its nihilism, is Ghost (Tatiana Maslany), a girl Brigitte befriends in [a] care facility. The bond Brigitte develops with this fellow 'little freak' (according to one girl who viciously pulls Ghost's hair) is intense. Yet ultimately Ghost betrays Brigitte. She locks her up in a basement just as her transformation into wolf completes itself. Any bonding, then, is virtually non-existent. [...] The betrayal by Ghost, encapsulated in *Unleashed*'s last line, 'and so began Ghost's reign of moral terror', reaffirms *Ginger Snaps*' cruelty.
>
> (Mathijs 2013: 119)

Mathijs notes that Fawcett 'also touched themes familiar to *Ginger Snaps* through his direction for the Canadian-shot television series *Lost Girl* (2010–), *The Secret Circle* (2011–) and *Orphan Black* (2013); the first concentrates on a succubus, the second is a show set in a high school about six teenage witches who form a coven' (2013: 122–23). A fascination with female bonding, and disruptive otherness challenging conventional (patriarchal) social norms, characterizes each of these Fawcett projects.

A coldly sadistic vision marks the fictional totalitarian societies created by Wells's Doctor Moreau and Huxley's Brave New Worlders, who are ruled by beliefs in survival of the fittest and domination, even extermination, of the weak. *Orphan Black* consciously revives this harsh dystopian vision in the oppressive institutions that act as antagonists to the Leda clones' freedom. These various institutions (the Dyad Institute, the Proletheans, Project Castor, Neolution) aim to instrumentalize and exploit the clones and any 'genetic material' linked to them (Kira, Helena's babies, Yanis). At the same time, *Orphan Black* revises the unmitigatedly severe worlds imagined by Wells and Huxley and their literary descendants like Levin and Wilhelm. It celebrates the existence and endurance and strength of affiliative families. (Think, for example, of the family dinner at Bubbles at the end of Season 3, and Alison's speech 'Thank you, sestras....') By emphasizing the differences among the clones, *Orphan Black* unambiguously valorizes nurture over nature. Ultimately, the inability to nurture, or the complete refusal to practice nurturing, is the greatest flaw in Westmorland, Coady, Moreau and the proponents of the Brave New World that Rachel Duncan hoped to rule. There are, in fact, many different kinds of family developed and explored by Sarah and her clone sisters throughout the series – biological, adoptive, affiliative and perhaps others still – that seem to speak not to domination and control, but rather to the emergence of a community.

Notes

1 See Otto Rank's *The Double* ([1925] 1971). Rank explores the double's uncanny otherness/sameness via a psychoanalytic approach.

2 The edition of *Evolution and Ethics* that I am using actually gives two different page numbers for each page of Huxley's essay, one indicating the page number of the whole volume (which is preceded by a lengthy introduction), and one indicating just the page of the essay. Hence, I give both sets of pages in parentheses after each quotation from his essay.

3 'The literal inspiration for Bailey Downs comes from Edmonton's Sherwood Park suburb, where [*Ginger Snaps* screenwriter Karen] Walton lived (the name Bailey Downs is a play on the city's south-central Bonnie Doon neighbourhood' (Mathijs 2013: 13).

Orphan Black and the Ideology of DNA

Hilary Neroni

The series *Orphan Black* provides an unexpected challenge to the ideology that surrounds DNA. This ideology, which today has an almost unquestioned hegemony, promises to eliminate the problem of the psyche and the disruptiveness of the unconscious that constitutes subjectivity. It eliminates this by fully mapping the body and brain in one deterministic system. The premise of *Orphan Black* is that a corporation has cloned one woman's DNA and created many women (all played by actor Tatiana Maslany), whom the corporation (the Dyad Institute) then placed as babies with families around the world and kept track of in order to study the evolution of this one form of DNA. (As the series unfolds it becomes apparent that the project also has other purposes, some of which are less scientific and more personal to the creators of the project.) The series follows one of these women in particular, Sarah Manning, discovering a tangled web of corporate and scientific intrigue as Sarah, a free-thinking foster child, unwittingly bumps into several of her clones and uncovers this international cloning experiment. As Sarah discovers her clones, whom she eventually calls her sisters, the series reveals another important premise of the show – that though the clones share the same DNA, they have all turned out extremely different from one another in terms of their personalities, their desires and their interests. The key to the intervention that the series makes concerning the ideology surrounding DNA today lies in these differences between the female characters.[1] These many types of femininities highlight the potential for a plethora of subjectivities even though they all have the same DNA. The series explodes the ideology of DNA – the idea that DNA provides a signifying chain (a language or code) that is completely truthful.

The big reveal at the end of the first season, which Sarah and her sisters discover after researching their DNA, is that the Dyad Institute, the initial villain in the series, has actually legally patented their DNA. Each clone has a different DNA tag, which includes an identification code that says: 'This organism and derivative genetic material is restricted intellectual property', followed by a series of patent numbers. This illustration of corporate control over DNA provides the show's most concrete suggestion that the genetic is overlaid by political contestation. Political decisions – like a corporate copyright – underlie all the viewers' interaction with DNA in the series. The subsequent seasons of *Orphan Black* investigate this intersection of the law, consumerism, subjectivity and genetics. By revealing that DNA always emerges in a socioeconomic context, the series demonstrates how even scientific discoveries about our biology arise out of ideological moments that are rife with social conflict and unconscious decisions. Even more importantly, however, the series reveals that we cannot rely on the idea that biology is the ultimate resting place for the truth about an individual. It does this, not surprisingly, through a focus on femininity. By exploring the contradictions of female subjectivity, the reigning biopolitical conception of the individual (one defined solely by their biology) becomes visible, as does a rejection of the notion of DNA as truth.

DNA as Depicted in *Orphan Black*

Orphan Black takes as its point of departure several assumptions about DNA, assumptions that are culturally widespread – specifically, that DNA is the blueprint to our individuality. The opening credit sequence of the show, for example, provides an artistic elaboration that links the content of the show (the clones and the plight of Sarah Manning) to the DNA double helix. To the multilayered sounds of singing, sighing, female breathing and a combination of instruments and synthesizers that expand on the female breath, the abstract title sequence of cells dividing, plant-like entities multiplying and DNA strands crashing into each other ends with the words of the title *Orphan Black* all in capital letters. The stark letters on a white screen emphasize that the line connecting the 'H' is actually a DNA double helix. Initially, then, the series carefully links itself to our current cultural understanding of DNA as an essential marker of our individuality. It also suggests the logic that DNA as such makes identifying individuals much more concrete and scientific. As in other shows that incorporate DNA into their regular plot – such as *CSI* (2000–15) – the characters in *Orphan Black* often collect bodily samples to uncover DNA. The concern with DNA becomes clearest in the basic conceit of the series, which involves multiple clones of a single embryo gestated in many women around the world. But as even the abstract title sequence hints, the series does not just present DNA as an answer but reveals the ideological and unstable nature of the assumptions associated with it.

The premise of *Orphan Black* relies on our cultural fascination with the current scientific concentration on the potential role of DNA in explaining who we are. In recent decades, we have more and more come to embrace the idea that DNA is the true identifier of an individual. The science behind DNA, for example, makes fingerprinting as a method for identifying an individual seem quaint. Over the last century, we can trace our investment in the biological identifier from the first systematic use of the fingerprint as a biological identity marker by police in New York City in the early 1900s to today's belief that DNA is a foolproof identifier, an identifier that tells us the truth about an individual. As Giorgio Agamben argues, 'contemporary citizens, lost in an anonymous mass and reduced to the level of potential criminals, are defined by nothing other than their biometric data and, ultimately – by means of a sort of ancient fate, which has become all the more opaque and incomprehensible – their DNA' (2011: 52). Agamben recognizes that DNA has paradoxically made individuals appear transparent while at the same time introducing a total opacity. This move from not relying on biological identifiers to expecting each person to have a foolproof biological identifier is revelatory. The creation of a perfect biological identifier is an expression of the twenty-first century and the ideology that characterizes this epoch.

DNA as the truth of our individuality is just one expression of a turn in the twentieth and twenty-first centuries to an ideology of biopower. Biopower is an appealing ideology because it offers us an image of human individuality without the stain of subjectivity, the mark of unconscious desire. Biopower makes use of technological innovations to eliminate the subject from our calculations about the individual, as, for instance, when we explain alcoholism with recourse to the genetic make-up of the alcoholic. Of course, the gene and DNA do play a crucial role in determining the biological constitution of individuals. But the culture at large has latched on to the idea of DNA as an explanation in order to create an image of perfectly explicable individuals that do not act like subjects, who instead often act in ways that undermine their own good or their own biological dispositions. Contemporary society privileges DNA, as we can see in television shows, the medical industry, security systems and so on. By turning to DNA for answers, we slip into the ideology of biopolitics and attempt to avoid the problems of subjectivity.

History of Genetics

In many ways, genetics proper begins with Darwin's theory of natural selection and, more specifically, with his idea of heredity. In 1860, a year after the publication of *Origin of the Species*, Gregor Mendel, an Austrian monk studying the pea plant, proffered the original idea of genetic inheritance. But it wasn't until a trio of papers published in the *Proceedings of the German Botanical Society* on the subject in 1900 that genetics as a science emerged. The first half of the 20th century saw an early investigation into what even the word 'gene' might mean. It wasn't until 1943 that scientists identified DNA as the 'carrier of biological specificity in bacteria' (Keller 2000: 3). And in 1953, Watson and Crick famously discovered the double helix of DNA that formed genetic material. The advances speeded up considerably with the invention of more advanced DNA processing technology in the mid-1970s and became especially high profile with the launching of the Human Genome Project in 1990. the Human Genome Project, I argue, in many ways represents the ultimate biopolitical accomplishment.

The Human Genome Project was an attempt to sequence human DNA. DNA is made up of four specific chemicals that align themselves into multiple combinations of base pairs. The order of these pairs determines the information available for building and maintaining an organism. The Human Genome Project tried to identify the entire human DNA sequence. One early developer of sequencing technology and biochemist, Walter Gilbert, enthusiastically wrote, '[t]hree billion bases of sequence can be put on a single compact disc, and one will be able to pull a CD out of one's pocket and say, "Here is a human being, it's me!"' (quoted in Keller 2000: 6). Gilbert's language here is revelatory. The promise of genetic research for its enthusiasts is identification – a scientific measure indicating who one is.

The Human Genome Project accomplished its mission to much fanfare in 2003. Conveying the excitement of the achievement, people have compared the completion of this project to landing on the moon. But if we think about it in terms of ideological significance, landing on the moon pales in comparison.

What I want to argue here is that DNA as a concept and the mapping of human DNA, while certainly an incredibly important scientific discovery, is also an expression of the ideology of biopower, which reduces subjectivity to its strictly biological coordinates. Theorists such as Michel Foucault and Giorgio Agamben conceive of biopower as a societal structure in which our subjectivity is reduced to bare life.[2] These theorists see biopower as an inevitable evolution of the trajectory of historical strictures on the individual (from the medical industry, education, modes of punishment and so on) that construct a restricted subjectivity. My argument, however, categorizes biopower as an ideology rather than a totalizing social structure. That is, biopower is not how power actually operates today but rather a way that subjects deny struggling with their subjectivity. For instance, I see Zoloft as the ultimate solution to my depression rather than exploring the psychic or social causes of it.[3] Biopower is a way of justifying social relations, not the actual logic of social relations.

This key distinction enables us to avoid struggling against biopower on its own terrain, which is what happened in the case of Michel Foucault. Foucault struggled to articulate an alternative to biopower because he accepted its reduction of subjectivity to the body. Theorizing biopower as an ideology, however, acknowledges the role that unconscious desire plays in undermining the body's inherent pleasures and pains. Subjects live within biopower's strictures (in the constant monitoring of our body, in the emphasis of our physical survival over the identification of values, or in the constant biological explanation given for our behaviour) but also see the way that the biopolitical explanation falls short. For instance, people today see the surveillance of their bodies through such new technologies as the Fitbit or the Apple Watch as the way to ensure good health. The ideology of biopower leads us to believe that such constant monitoring of the biological will ensure our optimum state of existence. These monitors record your bodily functions and encourage your best behaviour. They tell you to move each hour, how many steps you should be taking, when to sleep and wake, and the number of calories to eat. What they can never monitor or quantify is the individual's desire and even their desire to sabotage their own progress. But when such a failure occurs, biopower tells us to be more vigilant of our body to overcome such dalliances. The ease with which DNA became quickly folded into the ideology of biopower shows how all-encompassing this ideology has become today.

Alongside the discovery of DNA as an identifier of individuality, researchers hoped the discovery of DNA would reduce human suffering. One of the promises that we often hear from scientists is that if we discover exactly which gene is mutated for which disease we can not only create cures or immunizations for conditions like Alzheimer's disease, and breast cancer, but also begin to shape our own individual evolution. That is to say, we can begin to choose our genetic characteristics either for our unborn children or potentially for ourselves. Even before this future possibility, we can at the very least now know if we have proclivities for deadly diseases. Famously, Angelina Jolie recently had a double mastectomy and complete hysterectomy because doctors told her that she had a mutated gene that most likely indicated breast cancer.

Not surprisingly, while the ideological idea of DNA as the one true marker of us as individuals has gripped the cultural imagination, scientists are actually beginning to move on. Once the human genome was sequenced they found some alarming things, namely that in some ways our DNA is not simply our DNA. Instead, it is always already falling apart, morphing, changing and acting unpredictably while being hijacked by viruses and bacteria. As James Watson himself would say after the mapping of the human genome, '[a]ll of us have about 10 places in our genome where we have lost a gene or gained a gene. We are all imperfect' (Watson 2005). Watson's comments reveal that within the early work to discover DNA was an idea of the perfect set of information that then was revealed to be imperfect. And in a book titled *The Century of the Gene* (2000), biologist Evelyn Keller claims that in 50 years biologists won't even be using the terms 'gene' or 'DNA' since they have already moved on to systems and enzymes that surround the DNA as primary sites. This idea of our core truth as a perfect code, as information, however, remains a key part of the ideological function that DNA performs.

The identification of the subject with its DNA appeals to us because it imagines a signifying chain, an internal code that explains everything about us without any fissures, uncertainties or any lack. According to this ideology, this belief in an internal code that holds the truth of our being, there is no point where knowledge breaks down and the subject's desire manifests itself. In other words, this ideology relies on subjects who are biologically rational, who would never engage in harmful or dangerous behaviour. Every subject in some small or large degree, however, feels compelled to do things that go against what they know they should be doing to keep their body in good health (from eating more calories than is ideal to excessive drinking to snorting cocaine). The ideology of biopower, however, posits a key written into the body itself that tells us exactly who we are and how we should behave to keep our bodies at their best. In this way, the ideology of biopower takes up the dream of perfect knowledge.

The problem with this dream, as Jacques Lacan points out, is that an uninterrupted signifying chain cannot provide any sense. According to Lacan, sense depends on a point of nonsense, a point that interrupts signification. This point of nonsense is what Lacan calls the master signifier, a signifier that has no signified meaning. The master signifier interrupts the chain of signification because it doesn't have a signification, and yet, it provides the foundation for all significations. For instance, monotheistic societies treat God as the master signifier. In the last instance, God explains everything, but nothing can explain God. God, as a master signifier, is a signifier without a signified. Every system of signification has such a signifier that grounds sense and yet is nonsensical. The promise of constructing a system of signification without a point of nonsense can never be realized.

This is the ideological fantasy of DNA. Even the image of DNA, the double helix that stretches on and on, promotes this idea of an endlessly self-supported closed system. If we are able to eliminate the cut within the signifying chain, we can simultaneously eliminate loss, which is indicated by the interruption of the master signifier. This is why those who champion DNA research point to its ability to assist us in overcoming disease and even death. The ideological fantasy of DNA is that if we know the secret of DNA's signification without loss, we will learn the secret to transcending loss in our own lives.

But this image of DNA, like all imaginary structures, doesn't deliver on its promises. Instead, it provides loss for us in a new guise. The attempt to eliminate the necessity of loss, the repression of the necessary loss, inevitably inaugurates its return elsewhere.[4] The loss of Angelina Jolie's breasts would mark a concrete instance of the return of loss through the investment in the image of wholeness proffered by thinking oriented around DNA. The image of the perfectly mapped body ironically impels us towards further sacrifices in the name of a world without sacrifice. One can't eliminate loss or fissures in the signifying chain without eliminating signification altogether. Certainly, most scientists began to understand this as they changed their way of thinking about DNA, but the culture at large latched onto this idea about DNA as a complete and unerring truth of the human individual.

Orphan Black takes this ideology of DNA as its point of departure. It has several characters that proffer the reigning idea of DNA. For example, in Season 1, Episode 6, Dr Leekie (Matt Frewer) says about his scientific corporation, 'Neolution gives us a choice as a self-directed evolution. I believe that's not only a choice but a human right'. In other words, in the ideology surrounding DNA, knowing and changing one's DNA is seen as a human right. In this sense, human rights have been turned into biological necessities rather than universal ideas such as freedom, justice and equality. But the series is careful to contradict Dr Leekie's position as well as it also makes clear that DNA does not provide the answer to the problem of subjectivity. Even though many characters on the show have the same DNA, their desires constantly conflict with each other and even with what is best for their own bodies. As a result, the 'Clone Club' that they form is always a tenuous affair. The series makes clear that the attempt to discover subjectivity through DNA ends up revealing excess that DNA cannot account for. And it is this excess that is the point of the spectator's enjoyment of the series. The series seduces viewers with the spectacle of perfect genetic duplication in order to reveal to them how this perfect reduplication inevitably ends up reproducing too much. It is in this *too much* that an alternative to the genetic future resides. The 'too much' in this series is the contradictory nature of femininity itself.

At first glance, *Orphan Black* seems only to suggest that nurture trumps nature. In Season 1, Episode 4, the characters explicitly discuss this question. Cosima says, '[w]ell, yeah, bad brain chemistry can be genetic, but environment – that's individual. That's the whole nature nurture question'. The viewer understands that each clone has turned out differently because she is raised in different families and nations. It quickly becomes clear, however, that simply turning to the explanation of nurture doesn't fully address the varied desires of Sarah and her sisters. Felix (Jordan Gavaris), Sarah's foster brother (and eventually adopted foster brother to all the sisters), says in response to Cosima's nature/nurture comment quoted above, 'okay, let's talk nature. Out of nine so far: one's a psycho, one committed suicide, one is a bloody soccer mom, and then there's you. My crazy sister is sane by comparison'. In other words, for him neither nurture nor nature could possibly explain the unconscious desires of any of his sisters. In this and many other ways, the series is constantly stressing their biology while highlighting the visibility of the differences in their cultural identity. Biology and culture don't work together to define the individual in *Orphan Black*; instead, it is the constant clash between them out of which the subject arises. Because the subject lives this conflict rather than knowing it, she necessarily has an unconscious. Due to its concept of the subject, the series highlights their different desires, which are reducible neither to biological identity nor to cultural difference.

The Clones of *Orphan Black*

Orphan Black represents society's current fears, anxieties and desires surrounding the discovery of genes and DNA in addition to biopower's idea that DNA is the ultimate truth of our being. In order to confront this idea, the series focuses on the multiple instances of female subjectivity that one DNA sequence produces. The decision to depict multiple women is theoretically significant. Female subjectivity is the site at which the contradictions of the signifying chain become clearest because this subjectivity, unlike the male version, has no anchoring point in the master signifier. There is no one female ideal but instead two conflicting ones – mother and sexual object. As a result, any effort to be the ideal woman is always at odds with itself unlike the effort to be the ideal man. For instance, women who take up leadership positions, especially as the head of a nation, are subject to these conflicting ideals of femininity. Society finds it difficult to accept that the leader of a nation would be a sexual object so women in this role – such as Margaret Thatcher or Angela Merkel – must conform to a more maternal or matronly image. Regardless of appearing sexually attractive or more fatherly or grandfatherly, male leaders do not have to contend with the same scrutiny since all these characteristics simply fall within the larger ideal of male power within patriarchy. For women leaders, however, they constantly have to navigate the opposing ideals of femininity. In this way, female subjectivity indicates the contradictions of the signifying chain within patriarchy in its unadulterated form.

The show engages the viewer with the multiple types of femininity that the different cloned women exhibit. Each of these character's traits derives from her unique subjectivity, which is expressed in various aesthetic ways. For example, each has a different accent, style and movement that defines her. Sarah is the central clone. She is a streetwise woman in her twenties brought up initially in England and then whose foster mother moved her to the United States. Some of the main clones she interacts with include Alison Hendrix, a suburban housewife and soccer mom; Cosima Niehaus, a microbiology Ph.D. student from California; Helena, a trained assassin raised in Ukraine and Rachel Duncan, a chilly high-ranking official within the Dyad Institute, the company Sarah discovers is behind the cloning efforts. She also interacts with other clones such as Tony, a savvy drifter, Katja, a

German woman, M. K., a reclusive hacker and Krystal, a hairdresser. When watching the series, one relates to these various incarnations of the same genetic material as completely distinct subjects.

The series has the same person, Tatiana Maslany, play all these different characters. The technology that the series employs aids her acting by allowing her to appear multiple times in one frame. Working with a body double, body sensors and a special 'TechnoDolly' – which repeats complex camera movements exactly – the series shoots these scenes multiple times to allow for these characters, played by the same actor, to speak to each other, dance with each other, and even hug and hit each other. Watching Maslany play all these roles is an often-cited component of the viewer's pleasure in watching *Orphan Black*. Many reviewers remark on Maslany's performance and describe in detail their admiration and pleasure when they remember that she is playing all the roles. As *Time*'s Graeme McMillan writes, '[t]he key to that connection isn't simply the writing, but the performances – and most important, the performance of Tatiana Maslany, the lead actress who manages to juggle ten different characters in a way that seems surprisingly believable' (McMillan 2014). Though viewers are well aware that actors play different types of characters with a wide range of gender characteristics throughout their careers (some more than others), it becomes a more concrete realization when the viewer experiences these characters all at once. The fact that the clones share their DNA with their sisters but remain completely unique anchors these different characters (and their differing gender and sexual desires), so that the viewer experiences each woman as singular rather than various degrees along a sliding scale of one DNA expression.

As the story of their cloning and their biological predicament unfolds, the antagonism between their subjectivity and their status as clones becomes apparent. Each woman has her own form of subjectivity – one defined by her way of relating to her subjection to the signifier – that cannot be explained by the DNA that she shares in common with her sister clones. This contrasts with the male counterpart clones that the series introduces in later seasons. Importantly, the male clones are alarmingly similar to each other. They do not have the same diversity of identities that the women do. This is explained by the fact that they were raised together for a specific military purpose, while the female clones were unaware of each other and brought up in different families and countries. Though this seems to rely on cultural determinacy to explain female difference, in fact the series spends little time on this explanation but instead consistently emphasizes that each woman is reducible neither to biology nor to culture.

In this way, the series stresses that individuality is not inborn. It derives from the contradictory nature of subjectivity, and the way we inhabit the contradiction defines our subjectivity. Subjects are always at odds with themselves because they necessarily relate to themselves through the Other. For the male subject, the path to self-definition lies through identification with an ideal figure that exists outside the signifying structure (God, the father, the hero, etc.). The male constantly confronts his lack in the face of this ideal. For the female subject, there are only options within this structure. Women must identify as a mother and a sexualized being, for instance; even those two positions contradict each other within contemporary ideology. Female subjectivity is thus inherently *excessive*, which

Orphan Black indicates through its multiple figures played by the same actor. Through the figures of Sarah and the other clones, we see that what defines subjectivity is its excess. The excess of clones in the series stands in for the excess of subjectivity itself.

In the much-discussed dinner scene at the end of Season 3, for example, many of the surviving clones get together for dinner along with their close friends. They toast their deceased sister Beth (who died at the very beginning of the series). Helena stands to give her own toast at one point, and this brings Alison, Sarah and Cosima to their feet as well. All in one frame, the four women lean in and clink glasses. Clearly, the series with this one frame emphasizes its technical prowess (see Figure 15). At the same time, however, viewers, while knowing we are watching the technical wizardry of the series, confront the uniqueness of each of the characters. Alison gives a formal speech in which she says, 'thank you, sisters, for your bravery, for protecting us, for making us feel normal, and I know whatever comes next we will face it together as a family'. Here, the dialogue emphasizes the biological connection that they have with each other, but throughout this scene there is a clear conflict between the dialogue and the image. The dialogue stresses biological sameness while the image shows radical cultural difference.

This is one moment in the series in which the viewer marvels at the technology and acting skills while we watch all these clones appear on one screen and witness them interact with each other. They have interacted throughout the series, so that is not new – but seeing them all at once is exceptional. The image emphasizes their cultural differences. Helena, for example,

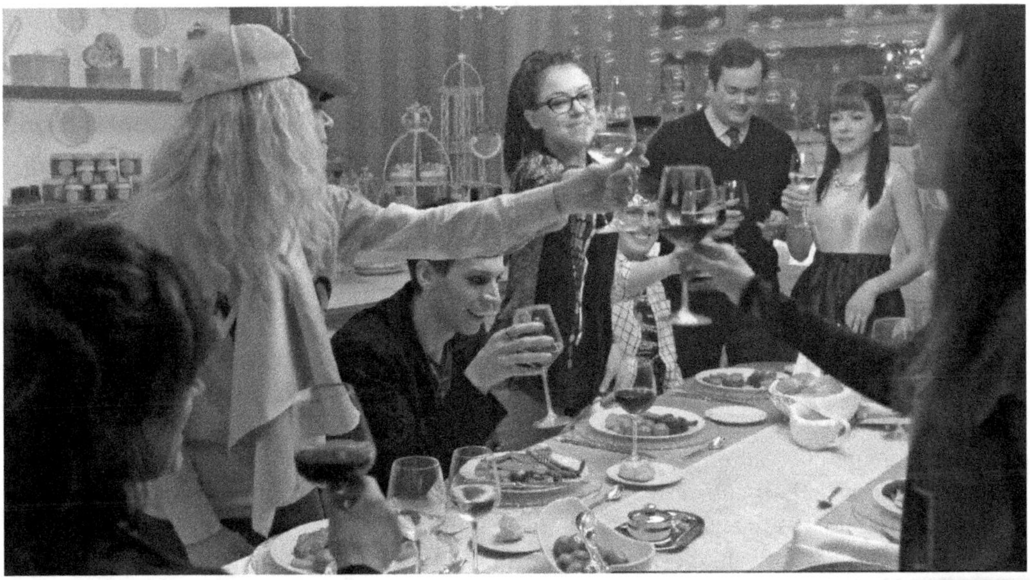

Figure 15: All four principal clones (from foreground to background, Sarah, Helena, Cosima and Alison) clink glasses in 3.10.

characteristically sticks out in her pink shirt, old baseball cap, massive blond curls and less than polite ways of eating, which contrasts tremendously with Cosima who is dressed in a chic bohemian manner with her dreadlocks pulled up. She uses her hands expressively and delicately to talk about some aspect of scientific discovery. Sarah is characteristically dressed in all black, her voice is slightly deeper to match her streetwise, tomboy past, and she emotes the sadness she feels for those who didn't make it this far. Meanwhile, Alison is dressed in a fancy dress, wearing make-up, and refusing to swear even at moments of great emotion. The confluence of the dialogue emphasizing their biological links and the image, which in turn emphasizes their cultural differences, makes the point on the level of form that it is neither nature nor nurture that prevails. Instead, the collision creates each character's singularity.

Orphan Black takes up the ideological conceptions of femininity and explodes them. By revealing to us so many positions of femininity – such as heterosexual women (Alison and Rachel), lesbian women (Cosima), transgender men (Tony) and wild and not sexually defined women (Helena), all of whom have varying degrees of maternal expressions and all played by one actor – the series reveals the contradiction in the usual opposing positions of femininity. Rather than endlessly shifting identities, the series reveals a plethora of singular feminine positions that together expose something quite different. These feminine positions point to the true nature of subjectivity – that it is stained by something that is too much. Rather than being completed biological individuals, these women – when the viewer experiences them throughout the duration of the series – bring to light the excess of subjectivity, which cannot be biologically mapped or quantified.

DNA, however, is a story of complementary base pairs that form a whole and identify the truth of individuality, even to the extent that it can predict that individual's future. There is little room here for the contradictory nature of subjectivity or the desiring subject who does not know her own desire. The show's intervention into this reigning ideology of DNA takes into account not just femininity and subjectivity but also the way that capitalism is bound up with the ideology of biopower. As I mentioned earlier, Season 1 of *Orphan Black* ends with the discovery that each clone has a tag in their DNA that says that they are 'restricted intellectual property' and have been given a patent number. If DNA is the truth of our individuality, then there is something there, and we can finally know who we are. We might not understand why the universe exists, but we can identify a concrete road map to our existence in our DNA. This belief in a material explanation of our individuality challenges previous assumptions about our existence. The turn of the twentieth century was marked by a confrontation with uncertainty or a new de-centeredness. With Freud's discovery of the unconscious, Nietzsche's announcement of the death of God and Marx's revelation of our lives being shaped by economic forces beyond our control, people no longer felt they had direct control over their subjectivity. The prevailing philosophies of the twentieth century stress the problem of subjectivity as the starting point for making sense of existence.

The turn of the twenty-first century, however, reveals a new reigning modality informed by an ideology of biopower. It is an ideology that re-centres the individual through the material body and through the effacement of the stain of subjectivity. In other words, at the turn of the twentieth century we discovered nothing and our relation to it, but at the turn of the twenty-first century we discovered something. This something is embodied in DNA itself, which can finally reveal the truth about who we are – including our past, our future and even how we will die.

But it is a truth that we can't access without the technological apparatus of medicine and science. It is also a truth we might be able to change (to make ourselves more resistant to illness or take away the cancer we are going to get), but we can't do anything about it on our own. We are reliant on medicine, science, health insurance and money to make this happen. This way of thinking about the truth of the individual is also ready-made for capitalism. The ideology of biopower is indeed the contemporary form of capitalist ideology. The more we conceive ourselves as merely living beings, the more we accede to the demands of a capitalist subjectivity.

Orphan Black addresses this when it imagines a powerful company (well-placed in the corporate and government structures), the Dyad Institute (and its scientific arm, Neolution), that has engineered the clone trials as well as created a way for parents to buy traits for their children and tweak genetic deficiencies.[5] For the latter service, they have set up a luxury clinic that administers their BrightBorn treatment. BrightBorn's slogan is 'Making the world better, one baby at a time'. This aspect of the ideology of biopower becomes central to the series in Season 4, when the investigation of Sarah and her sisters into the Dyad Institute leads them to a BrightBorn clinic.

During this season, Sarah sends her sisters and friends to the clinic to investigate. In the several episodes in which they try to infiltrate the clinic posing as potential clients, they discover that the clinic is altering the DNA of embryos to eliminate weaknesses or unwanted traits. Initially, this seems like a gross perversion of medical ethics, but it moves into the realm of horror when Cosima discovers that they are also running secret experiments with the unused embryos. These embryos are implanted into surrogates – who are lured there with the promise of a large pay for the service – but who often end up losing their lives when the pregnancies go horribly wrong. Through this sequence, *Orphan Black* warns against the dangers of the ideology of biopower, but it also reveals female subjectivity as the privileged site where these dangers manifest themselves. Of course, this is not a new revelation. Cloning gone wrong and women used in genetic experiments is not a new subject in cinema and television. It can also be found in films like *Alien: Resurrection* (Jean-Pierre Jeunet, 1997) and television series such as *Battlestar Galactica* (2004–09), where the women's subjectivity coincides with genetic experimentation. These experiments so often target women because the ideology of biopower tries to solve the contradictions of subjectivity and these contradictions animate female subjectivity more explicitly than male.

In *Orphan Black*, the Dyad Institute believes that it owns Sarah and her sisters since it has made sure that the Institute's patent is actually attached to the women's DNA itself. The idea of a patent being attached to DNA provides another intriguing image, which can point to the way DNA is conceived. It suggests that the patent, and capitalism as such, relies on the idea that DNA is whole and has no cut or lack. In this way, it is the completeness of

DNA that allows it to be contained, tagged and controlled. A DNA sequence with lack and fissures that constantly changes would always be slipping out of capitalism's grasp. Though the actual science of DNA does not smoothly align itself with the functioning of capitalism, the ideology of biopower conceives it in such a way that it does.

The series, however, takes the side of the women as they struggle to reclaim their freedom from capitalist control. *Orphan Black*'s first couple of seasons depict several layers of control that the Dyad Institute has over the women. The most substantial is the patent itself, but additionally the institute has planted someone in their lives that can monitor them and allow the team of doctors to test them in secret (usually drugging them before they go to bed so that the team of doctors can take them and put them through tests in the night). Initially, much of the tension in the series revolves around the women figuring out that they have a monitor and beginning to evade or counter this person. When the women in Sarah's Clone Club realize what they are, who their monitors are, and work to disentangle themselves from the grasp of the Dyad Institute, the Institute classifies them as 'self-aware' and works to strike a deal with them. Alison, for example, who discovers she thinks is her monitor and decides to sign a legal agreement with the Institute in which she agrees to be tested and tracked. In return, the Institute will leave her alone to live her life 'unwatched'. In signing this document, Alison is acknowledging that she is indeed the property of the Institute because it owns her DNA. Ultimately, however, this lawful agreement falls apart as Cosima uncovers the patent attached to their DNA and relays to Alison that they will never be truly free with this in place.

Cloning itself in the series functions as a wrench in the perfect structure of DNA because it defies the idea that each person's DNA is unique. Throughout the show, the police are constantly being thrown off their investigation because the DNA they collect, which normally tells the truth, leads them to the wrong person multiple times. For example, in Season 1, Episode 7, the police take a DNA sample from a dead body they are investigating. To their surprise, the DNA matches the DNA of a killer they are seeking from another case (who is in fact Helena). The dead body and the killer from a different case are both Sarah's sisters and clones so their DNA registers as the same person for the police. DNA evidence in the case of the clones of *Orphan Black* is not only unhelpful, it is misleading. By showing the failure of DNA to define the subject at every turn, the series undermines the biopolitical ideology that subtends our faith in DNA. While investigations into our genetic material will undoubtedly provide incredible new insights into our biological beings, such investigations will never solve the contradiction of subjectivity. This is the lasting verdict of *Orphan Black*.

Notes

1 An investigation of online fan sites reveals that the show's fan base seems equally interested in the multiplicity of gender displayed in the series and the scientific speculations about the ethics and potential ramifications of cloning.

2 See Michel Foucault (1995) and Giorgio Agamben (2005).

3 Of course, ignoring the biological is not the point either since sometimes an SSRI can actually allow a patient to then explore the psychic or social causes of depression. Either extreme – attributing everything to the biological or attributing everything to the societal – suggests a comprehensible closed system that simply does not exist. We must instead always be dialectically stuck between the two.

4 This is one of Lacan's basic propositions. He explains, 'whatever is refused in the symbolic order, in the sense of *Verwerfung* [foreclosure], reappears in the real' (Lacan 1993: 13). The rejection of symbolic loss portends a real loss.

5 Dyad is probably an allusion to the company that held the patent for two specific genes (BRCA1 and BRCA2) called Myriad Genetics. On 13 June 2013, the Supreme Court ruled that Myriad Genetics could not own a patent on these two genes because they only discovered but didn't create them.

Being Together: Immunity and Community in *Orphan Black*

Jessica Tanner

'Bring us together, Sarah'. In a scene from *Orphan Black*'s Season 4 episode 'The Antisocialism of Sex' (4.7), Beth Childs speaks to Sarah Manning from beyond the grave, entreating the clone who once assumed her identity and took up her quest for the truth of Project Leda to resist following in her suicidal footsteps. This bender-fuelled vision brings the series to a pivotal juncture, one that confronts two modes of narrative progression: will Sarah join Beth in opting for death, reproducing the past and the show's originary narrative by closing herself off from her sisters? Or will she diverge from that plotline, communing instead with the *sestras* who live on? Haunted by the revenant past of her investigative forebear, Sarah nonetheless opts to dwell in the present and look to the future, endeavouring to bring together the clones who are her genetic identicals and the series' multiple protagonists. In forestalling Sarah's assimilation to Beth's fate, the show opens onto a model of dispersed subjectivity whereby the Leda clones are united by, but not reduced to, the throughline of Sarah's narrative arc (and by Tatiana Maslany's masterful performance).

The pilot of *Orphan Black* sets the stage for Sarah's formation: the chance encounter with Beth's death thrusts her into a plot of social integration, as the antisocial grifter 'becomes' a cop and starts to connect with her estranged family and newly discovered sisters. The structure of that formation and socialization is reminiscent of the nineteenth-century *Bildungsroman*, which as Marc Redfield describes, 'narrates the acculturation of a self – the integration of a particular "I" into the general subjectivity of a community, and thus, finally, into the universal subjectivity of humanity' (1996: 38). And yet it is not Sarah, but her opponent – the uncannily persistent Neolution founder P. T. Westmorland, purporting to have been born in 1843 and to have hobnobbed with the likes of Charles Darwin and Arthur Conan Doyle – who ultimately takes up the mantle of that developmental model. Westmorland yokes his individual quest for immortality to a collective vision for humanity that involves (in the words of Ferdinand Chevalier) 'point-manning the genetic future of the human race' by isolating and replicating a genetic 'fountain of youth' (5.2). Ostensibly altruistic, Westmorland's plot instead reflects a hypertrophic individualism, a sovereign self-determination that seeks to implement 'self-directed evolution' on species scale. *Orphan Black* critiques that masculinist individualism by entwining Sarah's protagonism with that of her four primary *sestras* and assorted secondary Ledas, offering a vision of collective subjectivity that reconfigures traditional paradigms of individual (male) subjectivation and social integration. *Orphan Black* is a narrative of formation and deformation in tandem, a narrative of the creation of a community of *sestras* that profoundly complicates both the primacy of Sarah's protagonism and clichéd notions of solidarity as sisterhood.

Drawing on the work of Donna Haraway, Roberto Esposito and other contemporary thinkers of immunity, community and biopolitics, this essay argues that *Orphan Black*'s 'Clone Club' stages a form of community that is both non-appropriative and non-incorporative, one that counters the negative immunitary logic of Neolution and instantiates what Esposito has called an 'affirmative biopolitics' (2008: 194). Sarah's bio-narrative immunity – symbolically, a non-reactive tolerance of difference, as she (like Helena) lacks the protein upon which the pathogenic prion acts in the other clones – allows for a protagonism that is synchronically and identitarily distributed amongst the *sestras*, calling into question the rigid distinction between self and other that subtends traditional conceptions of identity as proprietary. In a mirror inversion of Rachel's sinister plan to 'marry the best of both worlds' (4.10) by joining bots and clones, merging genetic modification with a misguided sense of evolutionary determinism, Sarah's plot rejects consolidation in favour of a non-reductive model of communion, symbolically dispersing her fertile immunity into a community based not on a shared identity, but rather on a radically common (auto)immunity.

Immunity and Community

In recent decades, thinkers such as Haraway, Esposito, Jacques Derrida, Peter Sloterdijk and Jean-Luc Nancy have identified immunity as one of the fundamental paradigms of contemporary biopolitics and organizing metaphors of modern life – as Esposito puts it, 'the coagulating point, both real and symbolic, of the entire contemporary experience' (Esposito 2006: 51). Originally a juridical term referring to the exemption of an individual from the common law, immunity was appropriated as a biomedical concept in the late nineteenth century, denoting the body's capacity to protect itself from the invasion of pathogenic microorganisms (Cohen 2009: 3). As Ed Cohen has argued, this modern notion of 'immunity-as-defence' is a 'biopolitical hybrid'; in circumscribing the body as 'a defensible interior which needs to protect itself ceaselessly from a hostile exterior', immunity both 'construes the individual as a natural unit' and naturalizes a military mode of understanding the relationship between the body and its context (Cohen 2009: 14, 20, 274). It is this combination of protective and identificatory functions – the defensive delineation of self against the Other, via the protection of the body from external antigens – that makes immunity a multivalent metaphor for understanding contemporary society. Indeed, for Haraway, from whose work the episode titles of *Orphan Black*'s fourth season are drawn, the immune system 'is an elaborate icon for principal systems of symbolic and material "difference" in late capitalism', 'a map drawn to guide recognition and misrecognition of self and other in the dialectics of Western biopolitics' (Haraway 1991: 204).

The immune system's symbolic capacity for figuring the border regulation of individual and communal bodies – its role as 'the diaphragm through which difference, as such, engages and traverses us', as well as its semantic linking of life and law – has likewise led both Esposito and Derrida to explore the inexorable but often antinomic relationship between immunity and community in modern society (Esposito 2006: 54). Drawing upon the shared root of the affirmative *communitas* and the negative *immunitas*, *munus* (a reciprocal 'gift', 'duty', 'law' or 'obligation'), Esposito argues in *Bios* that '[i]f *communitas* is that relation, which in binding its members to an obligation of reciprocal donation, jeopardizes individual identity, *immunitas* is the condition of dispensation from such an obligation and therefore the defense against the expropriating features of *communitas*' (Esposito 2008: 50). Subtending Esposito's analysis is Michel Foucault's distinction between the classical regime of sovereignty, characterized by the power to 'take life or let live', and modern governmentality, a regime of biopower or the power to 'make live and to let die' (Foucault 2003: 241). Under the sovereign regime, according to Esposito, an artificial individualization – one that 'unites individuals to sovereignty by separating them' – is the immunitary mechanism that serves to ensure order and protect life against both external contamination and the expropriating force of community itself (Esposito 2008: 63). In the modern biopolitical context, the proprietary logic of immunization – the emphasis on what is 'one's own', and thus by definition 'not common' or immune – intensifies, as the immunitary mechanism comes to 'adhere to' or be contained within the bodies of individuals themselves (Esposito 2008: 63). Immunization is revealed to be a *pharmakon*: 'the intrinsically antinomic mode by which life preserves itself through power', it initially protects but eventually negates life (Esposito 2008: 46). While an immune system is necessary, since 'life, be it single or common, would die without an immunitary apparatus', excessive immunization closes off both individual and collective bodies to all otherness, precluding the freedom and the community – the very life – it was designed to protect (Esposito 2006: 53).

In Esposito's reading, it is this dialectical immunitary mechanism that explains an aporia in Foucault's understanding of biopolitics, that of how an ostensible 'politics of life' so often devolves into a thanatopolitical 'work of death' (Esposito 2008: 8). For both Esposito and Derrida, the hypertrophic immunity that is characteristic of modern life begets a pathological autoimmunity, 'an immunity that is destined to destroy itself together with the other' (Esposito 2006: 55). Both thinkers, however, allow for the possibility of an affirmative turn that would impede the progression of biopolitics into thanatopolitics. For Esposito, an affirmative biopolitics would involve rejecting the notion of the proper, abrading the immunitary logic of self-defensive exclusion and reconceptualizing community as an improper 'being in common' of singularities, a radically inclusive 'exteriorization of existence' (Esposito 2012: 44). Esposito locates this possibility in a revised vision of how biological immunity functions, as 'common immunity': immunity is not simply a militaristic defence against foreign invasion, he writes, but a circumscribed hospitality to otherness, as with the maternal body's immune tolerance of the foetus (Esposito 2011: 167). Immunity is not only a mechanism of separation, but of relation. This understanding of the body as 'a functioning construct that is open to continuous exchange with its surrounding environment' no longer conceives of immunity as antithetical to community, but as a fundamental enmeshment of self and other – a sign of the very possibility of a life in common (Esposito 2011: 17).

Whereas Esposito finds the potential for community in immunity itself and figures autoimmunity as the negative, self-destructive excess of immunitary defence, it is precisely in the discourse of autoimmunity that Derrida locates a means of breaking out of immunitary thanatopolitics and sustaining the openness of community: 'Autoimmunity is not an absolute ill or evil. It enables an exposure to the other, to *what* and to *who* comes – which means that it must remain incalculable. Without autoimmunity, with absolute immunity, nothing would ever happen or arrive; we would no longer wait, await, or expect, no longer expect one another, or expect any event' (Derrida 2005: 152). For Derrida, autoimmunity is a double bind. As a misrecognition and intolerance of self, it constitutes a threat to identity, but it is precisely in undermining the immunitary fiction of a 'proper' individual or collective identity – in 'protecting itself against its own self-protection' – that autoimmunity has the potential to short-circuit the progression of community into a pathological, self-destructive communitarianism (Derrida 1998: 73n27). By weakening the immune system that defends the self from the Other, autoimmunity prevents the foreclosing of vital difference: 'Self-contesting attestation keeps the autoimmune community alive, which is to say open to something other and more than itself: the other, the future, death, freedom, the coming or love of the other' (Derrida 1998: 51). While their terminology differs, both Derrida and Esposito find the potential for a politics of life (rather than a politics over life) in the (auto)immune dismantling of the notion of a closed, sovereign self.

With its baseline interrogation of the relationship between self and other, of property and agency and of biomedical and symbolic (auto)immunity, *Orphan Black* creates a space for the potential realization of Esposito's notion of affirmative community through its enactment of the stakes of the biopolitics of immunity and community. On multiple levels, the immunitary paradigm conditions the show's plot and its characters: Sarah's protagonism is predicated upon her immunity to the prion disease that threatens her fellow Ledas and afflicts the male Castor clones with an ultimately fatal cerebral glitch; it is the same immunity that makes her and Helena fertile, allowing for the creation of Kira and the 'miracle babies' that propel the Neolutionist plot; the disease that attacks the Ledas is an autoimmune disorder, a failure of the clones' bodies to recognize and tolerate some part of themselves. Beyond the immunity *topos*, the show's persistent identity play in the relationships between the clones constitutes a sophisticated reflection on individuality, relationality and the tolerance of difference: when Sarah encounters Beth as she throws herself in front of a train in the series' opening scene, for instance, it is simultaneously a formative encounter of self with self and of self with other. Pursuing these threads in tandem reveals that in the battle between the *sestras* and the Neolutionists, the show confronts two modes of immunity and community: a communitarian plot based on a negative and ultimately self-negating immunitary logic, and an affirmative community of partially expropriated, interdependent 'singularities', where immunity and autoimmunity cooperate through mutual (and self-) contestation.

Immunizing Humanity: Neolution and Its Avatars

'Neolution, a philosophy of today for tomorrow. Rooted in our past, in the evolution of the human organism': despite the contemporary setting and science, *Orphan Black*'s Neolutionist plot is fundamentally anchored in the nineteenth century, as Aldous Leekie's proselytizing speech suggests ('Variations Under Domestication' [1.6]). In this light, it is perhaps unsurprising that it is Victorian-era scientist P. T. Westmorland who surfaces in Season 4 (and in the twenty-first century) as the show's primary antagonist. For though Sarah and her *sestras* are the protagonists of *Orphan Black* – and though on the surface, Neolution's is a social, rather than individual, plot – it is Westmorland and his avatars at the head of Dyad, Topside and BrightBorn who take up the mantle of a nineteenth-century model of protagonism based on representative individualism. At over 170 years old (or so holds the myth that allows the man formerly known as John Patrick Mathieson to amass believers in his eugenic cause), and in possession of untold power, the shadowy Westmorland hardly fits the mould of the realist *Bildungsroman*'s exemplary young hero seeking to learn the ways of the world and become integrated into society. And yet, the show implicitly evokes that paradigm by positioning the Neolutionist and his followers as a negative exemplar of immunized subjectivity, one that throws into relief the affirmative biopolitical stakes of the Ledas' pluralist protagonism. Westmorland has a dual mission: his personal quest for immortality, and a broader drive, as Virginia Coady describes it to Sarah and Mrs S, 'to control human genetics and modify the species' by eliminating variation ('Let the Children and Childbearers Toil' [5.4]). This representative joining of singular and collective destinies is characteristic of the *Bildungsroman*, where, as Redfield observes, 'the temporal arc of the individual subject's *Bildung* always at least potentially exemplifies that of humanity' (Redfield 1996: 171). My contention is that this Neolutionist protagonism, which simultaneously sets up an individual narrative as an allegory for a collective becoming and subsumes the many under the one under the guise of individual progress, operates according to the negative logic of immunity that Esposito associates with the model of sovereignty. What appears to be a developmental narrative for the individuals involved and, by extension, for collective humanity is revealed to be an anti-developmental narrative for community; the quest to find the genetic 'fountain of youth', and thus to develop the human species beyond development, represents a definitive foreclosure of difference, an autoimmune communitarianism on species scale.

As Westmorland attempts to rewrite the human genome, he harnesses the scientific and corporate prowess of three women in turn – Susan Duncan, Virginia Coady and Rachel Duncan – all of whom believe, at least provisionally, that the collaboration with Westmorland serves their individual designs. As a Leda clone, Rachel, the 'corporate-raised psychopath' that Westmorland empowers to stand in his stead as the public face of Neolution, remains Sarah's primary foil (5.4). Her narrative trajectory is irrevocably (if obversely) tied to Sarah's and the other *sestras*'; that the show's clone protagonists and antagonist are genetically identical makes their respective fates a de facto statement on agency, identity, difference and community. Unique among the Ledas in having been raised 'self-aware', Rachel does not share the *sestras*' experience of 'becoming' a clone, or of discovering a genetic or affective affinity with the women who share her face. As she says to Paul early in the series, 'I've been self-aware since I was a child. I'm not exempt from the program. I simply enjoy a unique vantage, one with privilege' ('Ipsa Scientia Potestas Est' [2.5]). Rachel's conviction of her own privilege atomizes her; convinced that her 'self-awareness' affords her a sense of self the other Ledas lack, she rejects the community of her sisters in favour of an individual, and individualizing, quest:

Rachel:	Mr. Westmorland has made me feel whole. My whole life I've searched for my purpose.
Ferdinand:	You mean point-manning the genetic future of the human race?
Rachel:	Yes. It's why I exist.

('Clutch of Greed' [5.2])

Like Westmorland, Rachel seeks subjectivation and social integration through domination. Though she understands her subjection to the Neolutionist's designs, her role as the head of Dyad endows her with a sense of sovereignty, according to her the identity and purpose that would distinguish her from her genetic identicals. As she stands before the citizens of Revival after meeting Westmorland in person for the first time (see Figure 16), she invokes the evolutionary discourse of selection to affirm her merit and inspire others to invest in her – and Westmorland's – sovereignty:

> I am humbled to have pierced the veil and to address you on behalf of Mr. Westmorland. Like you, I was selected. I know now his hand guided my entire life. I know Susan Duncan, my mother, once stood here as I am, appointed to move the future forward. And she did. She created me. It is time to be brave, to sacrifice. The fruits of nearly 200 years of Neolution science are now within our grasp. And we here shall drink from the fountain first.
>
> ('The Few Who Dare' [5.1])

The progressive verticality of Rachel's narrative as she preaches the corporate line is both demonstrative and productive of the external power that delineates her ostensibly cohesive self. In including her audience in the select 'we' who will first partake of the genetic 'fountain of youth', she calls upon their own desire for distinction in order to subject them to her plan – thereby reproducing the structure of a sovereignty that immunizes its power through a self-legitimating fiction of individualization.

Despite the collective (if ultimately self-serving) rhetoric she channels as Westmorland's public voice, Rachel's self-set mission throughout the series is to affirm her agency as a clone – and thus, to affirm that a clone can have agency. In her first confrontation with Rachel on the island after surviving her attempted murder, Susan Duncan attempts to

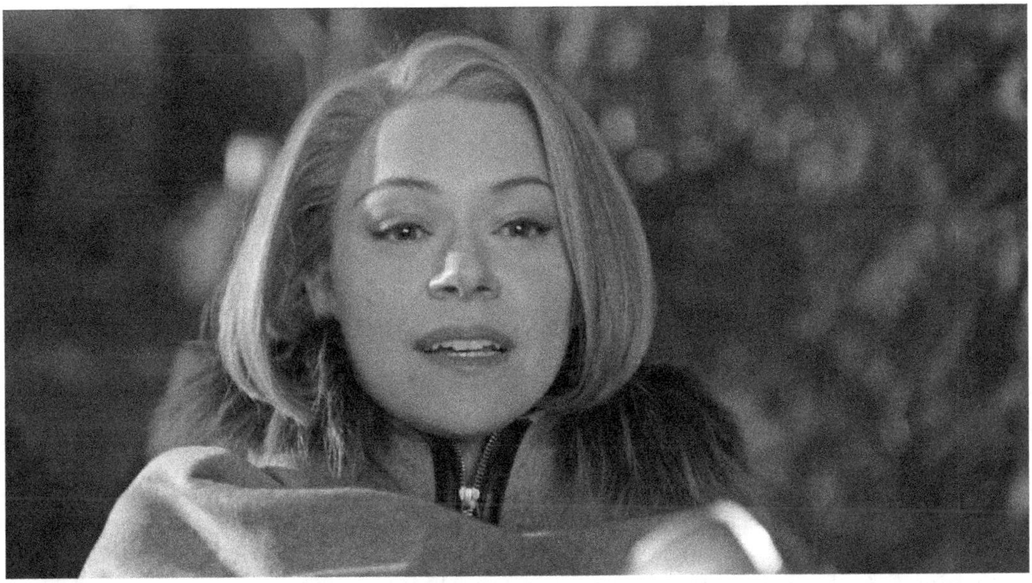

Figure 16: Rachel, 'appointed to move the future forward'.

persuade her daughter that they should collaborate, drawing upon their mutual interest in ensuring the future viability of the human cloning project: 'I want my life's work to go on for a very, very long time. That's you, Rachel'. Unfazed, despite her mother's reduction of the daughter she raised to the scientific advances her life represents, Rachel coldly replies: 'You didn't think I could accomplish this. [...] You said they would never respect me. That a clone couldn't have autonomy' (5.5). Conflating Susan and Evie Cho, the head of BrightBorn who once derisively pointed out that Neolution would never 'let a clone take a position of real importance', Rachel sets up her subjectivation as head of Dyad as allegory; she does not avow her desire for power in her own name alone (4.7). In reprising the language Evie used to dismiss her and designating herself as 'a clone', Rachel simultaneously distinguishes herself from other clones and assigns her individual autonomy – etymologically, a self (*autos*) law (*nomos*) – the power to signify their collective subjectivation.

This is, however, an equivocal autonomy; in practice, Rachel's ostensible self-determination (self-law) reflects an externally imposed politic of individuation (law of self). This is the false individualization of a sovereign regime that, to recall Esposito's analysis, presents individualism 'as the discovery and the implementation of the subject's autonomy' but deploys it as 'the immunitary ideologeme through which [it] implements the protection of life' (Esposito 2008: 60–61). It accords a negative, immunizing strain of freedom that is incompatible with community, Esposito argues, one that is 'understood as the mastery of the individual subject over himself – [...] his not being disposed to, or his not being at the disposition of others' (Esposito 2008: 71). Though she vaunts her position as a triumph over her fellow clones and those who doubted their humanity, the 'autonomy' Rachel attains is akin to the 'freedom' she offered Sarah in the first season finale: 'You'll remain free to live as you wish with Kira [...] unmonitored, but available. Call it "sovereignty association"' ('Endless Forms Most Beautiful' [1.10]).[1] By contracting herself to Westmorland, Rachel implicates herself in another form of 'sovereignty association', as she participates in the production and distribution of Neolution's power under the illusion of consolidating her own. In the age of biopolitics, as Esposito writes, the immunitary logic is disseminated and 'adheres' to individuals, 'as if the protective apparatus that is concentrated in the unitary figure of sovereignty is multiplied to the degree that sovereignty, once multiplied, is installed in biological organisms' (Esposito 2008: 63).

What Rachel seemingly fails to realize is that by operating within the proprietary domain of another – a system that quite literally owns her, as she carries the same proprietary sequence in her genome as her sisters – she remains subjected to its sovereignty as long as she seeks subjectivation on its terms. When Rachel attempts to put her mother in her place, boasting that 'P. T. put me at his right hand. Corporate runs the science', what goes unsaid is that both 'corporate' and 'science', both Rachel and Susan, remain subordinate to Westmorland and his shadowy institution, whose power is reinforced by Rachel's competitive drive for distinction (5.5). For despite the organizational squabbles between the 'corporate' Dyad and Neolutionist 'science', both (along with the military–industrial complex that drives Project Castor) participate in the strategic logic that Michel de Certeau identifies as 'the typical attitude of modern science, politics, and [the] military' (de Certeau 1984: 36). Strategy, in de Certeau's terms, is the self-segregating domain of the proper – of those with the power to claim a space as their own, to see and thus control those who inhabit it. Countless references in *Orphan Black*, both implicit and explicit, speak to Neolution's alignment with this strategic model of power: the monitors that subject the clones to surveillance and medical testing; Rachel's bionic eye, an implanted literalization of the surveillance whereby Westmorland not only sees her but also sees and controls what she sees; Felix's remark when he furtively looks out Mrs S's window and sees Rachel's associates that 'it's like we're in the bloody panopticon'; the isolated, proprietary 'island of Dr. Moreau' that serves as home base for Neolution's operations; the binary 'us' versus 'them' logic implicit in the name 'Dyad' itself (5.2). Susan, Rachel and Virginia Coady stand in synecdochically for the science, the corporate politics and the military respectively, but all three women, the series suggests (via Susan's death after refusing to capitulate to Westmorland's plans, Coady's forced killing of her last Castor 'son', Mark and Rachel's false sense of liberated empowerment), ultimately struggle to achieve true autonomy within Neolution. Westmorland makes use of women in order to achieve his aims; he assimilates them to his cause 'with fortune and fiction', as Susan observes, for 'that's how the patriarchy works' (5.6). Under his purview, their individuality is obscured, their differences strategically effaced so that they may be deployed for their use value and subsumed as symbol. Rachel's loyal embodiment of the brand, Cosima's science, Sarah's fertility and Kira's Lin28A mutation offer specific benefits to Westmorland's project, but it is the depersonalized collective of clones that he seeks to coalesce into a symbolic assertion of mastery over human variation – a mastery that the reprisal of human cloning would make infinitely and indefinitely replicable. The conquest of woman that frequently stands in as an allegory for individual and nationalist mastery in the novel of formation is reprised here on an expansive scale, as a conquest of the technology of reproduction that is generalized as a communitarian mastery of humanity writ large.

Despite Rachel's antagonistic opposition to the *sestras* throughout much of the series, the show elicits the viewer's empathy by pointing to the hermetic nature of her implication in the only community she has ever known. 'I'm not my sisters. I'm Neolution bred', she tells Evie when seeking a seat at the BrightBorn table (4.7). In a permutation of the logic of Helena's indoctrination by the Proletheans in Season 1, which weaponized her supposed distinction as the 'original' to eliminate the copies condemned as abominations, Rachel excludes herself from 'Clone Club' as a sign of superiority to the sisters she dismisses as indistinguishable. As Cosima astutely observes, in an emotional confrontation with Delphine, that immunizing separation is paradigmatic of Westmorland's strategy: 'This is what he does, he divides women' (5.5). Westmorland cultivates division, but refuses difference; he objectifies women (and men) but sustains the illusion of individuation in order to inoculate against resistance, to prevent the consolidation of a true community that would threaten his authority and proprietary interests. This strategy is perhaps most poignantly distilled in the series' penultimate episode, when Westmorland frames his command that Coady kill Mark and put an end to the Castor line of male clones as a sign of female subjectivation: 'Virginia, it's always been you and I with the strength to act. The future is female! Haven't you heard?' (5.9).

Beyond Cosima's mistrust of Delphine, and the infighting between the women Neolution empowers (Susan and Rachel, Susan and Coady, Rachel and Evie), it is Rachel's defensive isolation from her sisters that is the primary sign of the strategic division that empowers Westmorland. In the lead-up to the inevitable showdown between the *sestras* and the Neolutionists in its final season, *Orphan Black* sets the stage for Rachel's potential redemption through a dissociation from Neolution and a communion with her sisters – or her definitive isolation, if her alignment with Westmorland endures. It is only when Rachel realizes that she remains subjected to Westmorland's plot – that he is not who he says he is, that he is monitoring her through her bionic eye, and that her ostensible 'emancipation' from her monitors and her objectified identity as 779H41 is a ruse designed to ensure her compliance – that she opens herself up to the possibility of cooperation with her sisters (5.7). By seeing herself in Kira – an identification made manifest in flashbacks to the corporate childhood that formed her and her enduring alliance with Neolution's designs – Rachel breaches the sovereign isolation she has so carefully cultivated, as she is symbolically contaminated by the empathy that Kira embodies. Gouging Westmorland's eye out of her socket with a martini glass after returning Kira to Sarah, and further making herself vulnerable to her sisters and to Neolutionist attack by exchanging the information that will bring the corporation down, Rachel forsakes her immunitary 'autonomy' in favour of a freedom attainable only through interdependence and cooperation. And yet, Felix's admonition in the final episode as Rachel hands over the list of Ledas – 'Rachel, you know you can't come in' – affirms an estrangement from her genetic identicals that persists after her putative redemption, demonstrating through counterexample that the '*sestra*hood' is not a normative sisterhood founded on a shared identity or a proprietary immunity (5.10). The bounded plot of individual subjectivation achieved through redemptive social integration belongs to the patriarchal narrative logic that Westmorland perpetuates (and that perpetuates his myth) – a logic that the series persistently contravenes. Despite Rachel's ultimate exclusion from 'Clone Club', her provisional cooperation with the *sestras* stands as proof that agency and affirmative community can coexist, even – and perhaps especially – in the face of vulnerability.

(Be)coming Together: '*Sestra*hood', Clone Club and Common Immunity

In the trailer released in anticipation of *Orphan Black*'s series premiere, a sequence of close-ups showing Maslany portraying each of five clones introduced in the series' first season – Beth, Sarah, Alison, Cosima and Katja – culminates in a question: 'When did I become us?' (BBC America 2012). Referencing Sarah's admission into 'Clone Club', and her defiant consternation upon discovering her genetic identicals, the trailer likewise points to another of the show's fundamental questions, that of how community is constituted: not only when, but how and along what lines, an 'I' becomes an 'us'. When Sarah first learns she is a clone, her reaction is defensive denial, a self-protective reaffirmation of her individuality:

> Felix: You've got to be kidding me. Clones?
> Sarah: They're not me. They're not. They're completely different people.
> Felix: Soccer mum Sarah. Dreadlock science geek Sarah. Arguably more attractive than the real Sarah…
> Sarah: I'm just me, okay? The song remains the same.
>
> ('Variation under Nature' [1.3])

Upon discovering that her husband Donnie is her monitor a few episodes later, Alison laments to Sarah (who is impersonating Alison): 'you're the only person I can talk to and you're just another version of me. […] I'm not even a real person' (1.6). The truth of the Ledas' relationship to each other, the show suggests, lies somewhere in between Sarah's atomizing individuation and Alison's self-negation. The central narrative arc of *Orphan Black* is devoted to how the Ledas become a collective without losing themselves, but also without becoming stuck in an illusory conception of a fixed, cohesive identity – how Sarah brings the clones together, as Beth would have it, but also how losing the illusion of their sovereignty keeps them from falling apart. Unlike the Castor clones, who were raised together and subjected to the service of a military–industrial plot, the Leda sisters come together and become together; their mutual, relational development stands in contradistinction to the monolithic genetic determinism of the Neolutionists.

Subtending the show's depiction of the *sestras* as an affirmative community is a fundamental interrogation of the relationship between the body and possession or property. From its first season, *Orphan Black* simultaneously insists on the distinctive personas of the Leda clones and persistently undermines the very notion of the proper, as the *sestras* come to terms with the all-consuming nature of Dyad's hold on their bodies and their lives. Their lack of self-possession is most acutely referenced in the synthetic genetic sequence that Cosima discovers at the end of the first season: 'This organism and derivative genetic material is restricted intellectual property'. Distraught upon realizing that the Ledas' bodies have been appropriated by Dyad, Cosima remarks that '[a]ny freedom they promise is bullshit. [...] We're property. Our bodies, our biology. Everything we are, everything we become, belongs to them' (1.10). Though the show overtly condemns this logic of possession and women's lack of reproductive agency, it is nonetheless the putative dispossession of their bodies that brings the Leda sisters together – both literally, as they work together to discern Dyad's endgame and free themselves from its technocapitalist control, and symbolically, as it undermines the logic of proprietary individuation that prevents the development of affirmative community. This is not to say that the *sestras* are divested of agency. Rather, in their subjection to Neolution's designs and claims, they operate from a different position and in a different mode than Rachel and her cohort, making use of what de Certeau designates as 'tactics':

> a *tactic* is a calculated action determined by the absence of a proper locus. No delimitation of an exteriority, then, provides it with the condition necessary for autonomy. The space of a tactic is the space of the other. Thus it must play on and with a terrain imposed on it and organized by the law of a foreign power.
>
> (de Certeau 1984: 36–37, original emphasis)

Tactics are the tools of the weak, those who do not determine their field and rules of operation but instead act by 'poaching' on the property of others: readers (as opposed to writers), walkers (as opposed to municipal authorities and urban planners). They are adaptive, open and responsive to changes in the environment. Tactics imply vulnerability, which Judith Butler has recently argued for understanding as 'a way of being exposed and agentic at the same time' (Butler 2016: 24). Whereas Rachel attempts to seize power strategically, identifying agency with 'sovereign modes of defensiveness', *Orphan Black* is the story of how Sarah and her *sestras* find agency precisely in their vulnerability, as they gradually become receptive to each other and their context as a constitutive part of their selves (Butler 2016: 25). Their antagonist is not only Westmorland or Rachel, but the strategic immunitary apparatus that Dyad represents.

Though Sarah is the most prominent of the Leda clones, her protagonism is dispersed amongst the *sestras* – not through exemplarity (an 'I' that stands in for an 'us'), but through a plural relationality (an 'I' that becomes an 'us', that is receptive to alterity). Despite sharing a face and a genome, the clones are radically distinct and distinguishable, as Maslany's accents, gestures and hairstyles convey the radically different lives these genetically identical women led before 'I' became 'us'. That distinction reflects what Everett Hamner characterizes as the series' focus on 'the epigenetic tension between genes and environment' (Hamner 2017). The assemblage of *Orphan Black*'s Leda clones allows for the developmental narrative of epigenetics to be figured synchronically, contravening the progressive, deterministic plot that the show associates with negative immunity and the strategic communities it creates (the Neolutionists, the Proletheans, the Castors). As Catherine Malabou has argued, epigenetics allows for an understanding of biological difference itself as plastic, potentially constituting the grounds for 'a biological resistance to the biopolitical' (Malabou 2016: 429). Controlling the human genome, as Westmorland sets out to do, cannot control for the unprogrammed variation that epigenetic development entails; to borrow de Certeau's terms, epigenetic variation is a tactical response. In this sense, the plurality of their protagonism is in itself a form of resistance to the immunitary mechanism that keeps Westmorland in power.

That the same prion is the source of the Castors' neurological glitching, the sexually transmitted infection they are designed to disseminate, and the Leda defect points to the series' dual logics of immunity. In the Castors, the prion defect is weaponized, for the purpose of sterilizing their sexual partners by propagating the misfolded protein; in impeding reproduction, that sterilization defensively forecloses plot, a sign of the closed and (as the glitching makes manifest) ultimately self-destructive logic of negative immunization. In the Ledas, the same disease is self-contained, affecting multiple organs of individual bodies but bringing them together, generating story as they search for answers and a cure. Their distributed protagonism is a function of both Sarah's productive fertility and her immunity to the prion disease that besets her sisters, as the narrative of her formation becomes enmeshed with that of the other Ledas and her functional immunity symbolically counters their self-destructive autoimmunity. Sarah's (and Helena's) immunity is a non-reactivity to self; uniquely, since she is a clone, it also implies a non-reactivity to otherness. Though she carries the prion that provokes a pathological autoimmune response in her sisters, she lacks the protein that it activates – a tolerance of difference where the logics of immunity and autoimmunity are mutually held in check. This immune tolerance recalls Haraway's positive conception of immunity, which anticipates Esposito's:

> Immunity can also be conceived in terms of shared specificities; of the semi-permeable self able to engage with others (human and non-human, inner and outer), but always with finite consequences; of situated possibilities and impossibilities of individuation and identification; and of partial fusions and dangers.
>
> (Haraway 1991: 225)

It is likewise Sarah's fertile immunity that allows the Ledas to self-sustain, by halting the progression of the autoimmune disease. Deprived of access to Kendall Malone's original

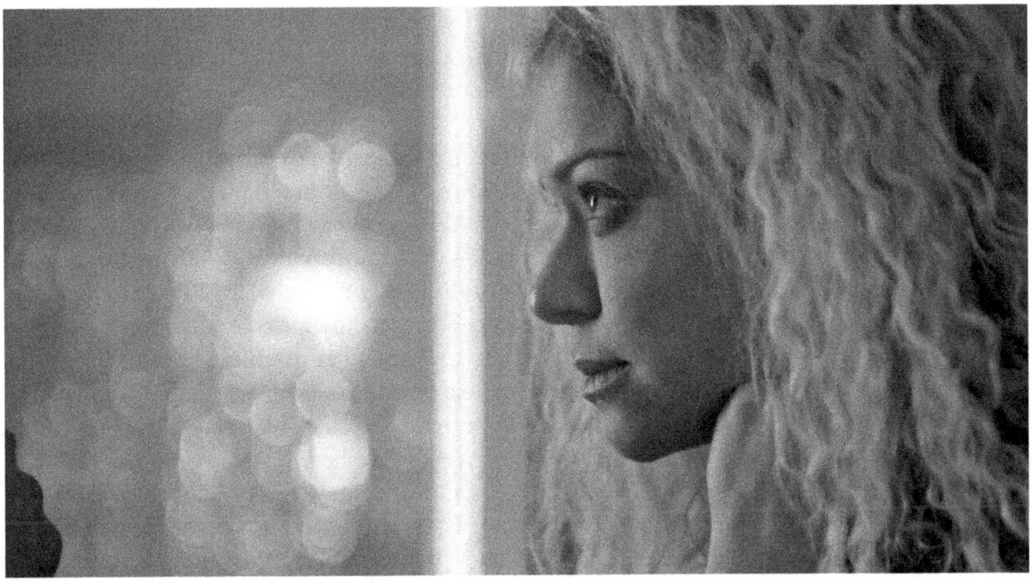

Figure 17: Helena recounts the lives of her *sestras* as her own.

genome, to a unity generated from and contained within one body, the *sestras* adapt to find a cure for the Ledas' autoimmune disease by tactically fusing Sarah's eggs and Ira's sperm – a cure that tellingly requires Cosima to develop what Westmorland calls 'a very elegant immunostealthing vector' (5.2).

The more obvious signs of the Ledas' permeable selves are the celebrated scenes in which the show layers the clones' identities, with Maslany deftly portraying clones impersonating other clones (occasionally impersonating yet other clones). In these scenes, they tactically adopt each other's identities in order to access information or escape danger, as Maslany's nuanced layering of the traits and quirks of distinct clones literalizes the dynamic implication with others that subtends the presentation of a cohesive self. Beyond these 'clone swaps', the sisters' self-presentation frequently reflects an imbrication of their narratives. In a scene from Season 2, Helena sits at a bar with an assortment of drinks before her, as Sarah researches Project Leda at the Cold River Institute. After Jesse intervenes during an unpleasant encounter with another patron, Helena offers him a White Russian along with her life story: 'In Ukraine, I was police detective. I shot many criminals. [...] Then I was brilliant scientist. But I quit to be with my family. [...] After rehab drinking problems. But now I am with my sestra, having adventures' (2.6). Jesse's response – 'You sure have led a life, huh' – belies the fact that this is, of course, a life that is not one. As Heather Duerre Humann observes, Helena's impromptu autobiography references details from her own life, but also from her sister's lives: Beth's police work, Cosima's science, Alison's stint in rehab (Humann

2016). The 'life' she claims as her own is plural; it presents the self (*autos*) as impressionable, entangled with the selves of her *sestras*. Well before Sarah manifests a receptiveness to the psychic connection with her mirror twin, Helena channels her sisters, letting them 'flow' through her as an integral part of her self.

Beyond the on-screen play of TechnoDolly-facilitated *sestra* gatherings and layered clone impersonations, the behind-the-scenes production of *Orphan Black* further contributes to the show's destabilization of proprietary identity. Critics often marvel at Maslany's 'shape-shifting', 'chameleonic', identity-bending prowess – at the ability of one actress to embody an ever-growing cadre of clones and make each of them distinctive (Elan 2014; Loofbourow 2015; Nussbaum 2014). This critical discourse echoes the show's first trailer: how does *she* become *them*? And yet, even Maslany's dazzling singularity is double. Noting that critics frequently cite the 'relational dynamics *between* the clones' as the chief evidence of Maslany's skill, media scholar Zoë Shacklock has recently pointed out that it is precisely these multi-clone scenes that rely on the unseen performance of Maslany's acting double, Kathryn Alexandre (Shacklock 2016: 70). Shacklock notes that the show only hides its constitutive multiplicity on-air; the paratextual material released as part of the show's promotional apparatus regularly features Alexandre, and goes behind the scenes to reveal how the apparent unity of multi-clone scenes is in fact a compilation of takes involving multiple actors, tennis balls and a TechnoDolly. In an interview, visual effects supervisor Geoff Scott reports that these scenes even occasionally require 'harvest[ing]' different body parts from Maslany and Alexandre and grafting them together to create a cohesive clone, a process he calls 'a little Frankenstein-y' (Prudom 2017). As Shacklock argues, *Orphan Black* 'explicitly problematiz[es] the singular, authentic self' by 'present[ing] performance as a multi-corporeal form of labor, formed through the patterns of presence and absence' (Shacklock 2016: 77). Even on the level of performance, the show unravels the fiction of an autonomous self by unveiling the ways it is sutured together to hide its relational dependence on others.

On-screen, this plural, dependent conception of self is most acutely realized in Kira's radical empathy, a psychic entanglement with the *sestras* that stems from a Neolution-targeted mutation on the Lin28A gene. As Kira puts it to Sarah in a scene from Season 4, 'your sisters, I know how they feel sometimes. [...] Like Cosima, when she's sad. Helena, when she's lonely. Rachel's the angriest. There's even some I don't know. [...] I feel you too, Mommy' (4.5). Kira communes with the clones across space and boundaries of familiarity: 'I can't feel her anymore', she wails to her mother when she senses M. K.'s demise under the force of Ferdinand's feet (5.2). While resulting from Susan Duncan's implantation of Janis's self-healing mutation in the Leda genome, Kira's self-regulating power to heal and to connect with Sarah's *sestras* reflects the futurity of Sarah's openness onto otherness. The synchronic dispersal of Sarah's reproductive capacity, her immune tolerance of otherness, is actualized in Kira's permeability. That Sarah progressively learns to 'feel' Kira and Helena 'flow through [her]' in the same way contravenes the logic of Westmorland's genetic determinism, as an epigenetic adaptation that stems from her connection with others (5.5). That tactical connection is inappropriable, as Sarah remarks to Helena, after reporting that she finally sensed the 'flow' of an entangled connection with her daughter and her twin: 'I don't give a shit about our genetics. This is what Rachel can't have' (5.4).

With Cosima on the island, Helena in the convent and Alison on a journey of self-discovery, Sarah's attentions in the early episodes of Season 5 are largely devoted to preventing Rachel from investigating Kira's mutation, a defensive immunization of the daughter she seeks to shield from Neolution's grasp. When bargaining with Sarah, Rachel remarks that 'anything we find will of course be proprietary. But she isn't. She's yours' ('Clutch of Greed' [5.2]). Kira is not however Sarah's property, as the show takes pains to show. Determined to discover the reasons for her difference, Kira initially rejects Sarah's efforts to protect her, refusing to comply with her mother's escape plan and cooperating with Rachel's testing program at Dyad. Beyond Kira's withdrawal, the first half of Season 5 decentres Sarah by presenting her inability to determine her *sestras*' actions, even in a protective mode: Cosima insists on staying at Revival despite Sarah's attempt to rescue her, resolved to 'follow the crazy science' (5.1); M. K. stays behind to face Ferdinand, wilfully exchanging her life for Sarah's freedom despite Sarah's protests and insistence that she continue to fight (5.2). A series of episodes in the final season devoted to each of the other primary clones, where Sarah's storyline is secondary, further reinforces her non-hierarchical enmeshment with her sisters.[2] In calling the primacy of Sarah's protagonism into question and depicting the futile nature of her efforts to protect her family by controlling their actions, the show rejects a breed of immunity that, though positively valenced in terms of the show's rooting interests, is incompatible with affirmative community. In accepting M. K.'s wishes and Kira's defiance, Sarah allows for the persistence of a form of freedom that is antinomic to Rachel's self-affirmed 'autonomy', one that Esposito characterizes as a disposition to others, 'that part of community that resists immunization […] and remains open to difference' (Esposito 2012: 56).

In her seminal 'A cyborg manifesto', Haraway considers what kind of politics and communities 'could embrace partial, contradictory, permanently unclosed constructions of personal and collective selves and still be faithful, effective' (1991: 157). The cyborg community that *Orphan Black* generates amongst the *sestras* responds indirectly to her question. A succinct vision of the dynamics of that community surfaces at the end of the promotional trailer for Season 5, 'The Final Trip':

Alison: Together, we are one.
Helena: We are one.
Cosima: We are …
Sarah: …one.

(BBC America 2017a)

Together, the *sestras* are, and are not, one, as this repeated, multivocal affirmation of their unity implies. They are, and are not, individuals; they are, and are not, an impersonal 'one'. '*Sestra*hood' is not a community based on an abstraction of the model of the individual, an allegorization of Sarah's formation. It is not a community of genetic identicals; as Beth says to her sister from beyond the grave, '[t]here's more than biology between us, Sarah. There's something else. You can feel it, too' (4.7). Neither is it a facile feminist assertion of sisterhood as an assemblage of women united by their common identity and experience (Felix is very much a 'brother-*sestra*', as Helena would have it). That it persists when '[t]here's no one left to fight', as Sarah observes in the finale, underscores the show's departure from the oppositional model of immunitary community, and its concomitant embrace (however tentative, on Sarah's part) of a 'freedom with', a being together, rather than a 'freedom from' a designated other (5.9). The community the Ledas progressively establish over the show's five seasons is based on a common (auto)immunity, on reciprocal obligation and implication, on entangled vulnerability. Tellingly, that community is not limited to the core diegetic Clone Club, nor even to the 274 other Ledas Cosima and Delphine set out to cure in the finale: *Orphan Black* itself is radically and indiscriminately open to communion, as evidenced by another #CloneClub, the online community of devoted fans that shaped the series over its run. Determining dialogue, props and Helena's food choices through their engagement with the show's official Twitter polls and Tumblr, #CloneClub also (re)created – and continues to create – *Orphan Black*, as BBC America encouraged fans to craft original fiction and art based on the series' characters (Koblin 2017). Like Helena's story of the same name, *Orphan Black* is 'an embroidery with many beginnings and no end' (5.10). This openness to intervention makes manifest the permeable, collective autonomy that the series simultaneously models and engenders, generating a communal futurity through the epigenetic re-visioning of the show the creators designed. As Haraway put it in a recent book, '[t]o be one is always to *become with* many' (2007: 4). *Orphan Black* is Sarah's story, but it is the story of her becoming (in) common.

Notes

1 The term evokes the Québec sovereignty movement, which asserted that the links between sovereignty and association were 'indissoluble'.

2 One of the show's writers, Alex Levine, reports that the show's creators had 'wanted to let clones other than Sarah drive the narrative since the beginning. But this year they realized each of the sisters is interesting enough that they don't have to be a capital "H" hero to keep us all engaged and keep the story driving forward' (BBC America 2017c).

The Dancing Women: Decoding Biopolitical Fantasy

Robert A. Rushing

Introduction

Biopolitical projects, from public service advertisements against drunk driving to billion-dollar stem-cell initiatives, are always animated and supported by a *fantasy*, generally that of the healthy, happy, safe and long-lived body. That body may be put at risk, or we might see the consequences to that body of a particular lifestyle. As Pasi Väliaho has shown (2014), certain images (what he calls 'biopolitical images') have a doubled temporality, showing both the present and an implicit future; think of an advertisement that shows a child holding a pack of cigarettes, or a driver with a beer in one hand and car keys in the other. Textual captions are almost superfluous in such cases – we can see the looming disaster, and long to intervene pre-emptively to prevent it. (Equally, the image might show the aftermath of the catastrophe, inviting us to imagine the moment when it could have been pre-empted.)[1] But while this animating fantasy is clear in the case of advertisements and public service campaigns, it is equally true that all biopolitical initiatives draw on a general image of a future in which catastrophe, illness, sterility and weakness are avoided – a body politic, and hence a nation, that is healthy and strong. At times, the cinematic and the televisual serve as the foundation for the politics of life and health, as when Reagan hoped to safeguard the nation against nuclear war with a 'Star Wars' initiative, or when Christopher Reeve and Michael J. Fox lent their celebrity status to supporting California's stem-cell research proposition. In every case, the basic biopolitical fantasy upholds the project: we can and indeed must securitize the future against the risks that are here now, whether we know about them or not.

At the same time, television and film often draw on biopolitical themes: melodramatic scenarios of families shattered by illness, or thrilling fantasies about disasters that threaten the whole human species, what Timothy Campbell calls 'species living' (meteor impacts, zombie menaces or other plagues, alien invasion). In general, the synergistic relationship between television, film and the biopolitical has not yet been fully explored or even recognized. Many of the staples of television drama, in particular, centre on life, health, vitality and security, including programmes about police, doctors and lawyers, not to mention the more overt cases, such as shows about forensic science or national security. The opening sequence of *ReGenesis* (2004–08), a Canadian series about a North American biological emergency response team (starring Peter Outerbridge, better known to *Orphan Black* fans as Henrik Johanssen), summons up just such anxiety-inducing biopolitical fantasies: we watch a man carrying an infectious plague touching doorknobs, escalator rails, money, a telephone, before he collapses in public. But that same opening sequence also showcases the fantasies of how science can 'securitize' those fears: images of pipettes, microscopes, cells dividing, needles and latex gloves. *Orphan Black*, of course, also relies on a series of 'biopolitical fantasies' that are fundamental to the show: secret human cloning projects; corporate, state, military and religious control of women's reproductive capacity; covert surveillance of our health, including our sexual lives; an infertility plague as a bioweapon – almost all of the show's primary plot lines centre on the use and abuse of power over bodies, life, reproduction and vitality. By Season 4, the show has even introduced wormlike artificial organisms, 'bots', that live in your cheek, monitor and experiment on your DNA and are fatal to remove – and Evie Cho, the CEO of BrightBorn, plans to introduce the bots into millions of women. These are fantasies, of course, in the sense that in the real world there is no biotech firm secretly implanting genetic robots into consumers, no real infertility plague. And yet, even the most naive viewer surely recognizes that *Orphan Black*'s fantasies are effectively more thrilling and melodramatic versions of that real world, or at least of the world as it might plausibly be in a few years. Indeed, as Gregory Pence notes, in the case of synthetic biotechnology, the real world is sometimes 'actually even more alarming than *Orphan Black* portrays' (2016: 87).

What is a clone to do in this world of weaponized infertility, bio-surveillance and appropriation of women's reproductive capacity? *Orphan Black* offered a range of possible responses for the viewer over its five seasons, but the powers arrayed against our clone *sestras* are formidable and mysterious: from ideologically driven transhumanist scientists to religious extremists, from neoliberal globalized biotech firms to the US military and intelligence agencies. Indeed, one of the most powerful suggestions of *Orphan Black* is that, while all bodies feel the application of biopower, it is not perhaps completely universal and neutral in its application to the human body. Some bodies feel it more than others, and indeed, it is the apparent universality of biopower that masks its uneven application: indeed, I would argue that, after watching a few seasons of *Orphan Black*, it is not really possible to think about biopower and biopolitics without also thinking about sex and gender. While all bodies are scrutinized and surveilled, and while everyone is pressured to behave in certain ways (what Foucault [1978] called 'discipline' rather than directly coercive force), some bodies are more scrutinized, more disciplined than others. Although *Orphan Black* does not extend its critique beyond sexual difference in equally visible ways (see, for example, Achille Mbembe's notion of 'necropolitics' [2013] – namely, that the relative health and longevity of the West is produced by exporting illness and death to the global South through pollution, war and cheap, unsafe labour practices), it definitely recognizes the differential impact of political power on men's and women's bodies.

In what follows, I want to look at one of the ways that *Orphan Black* has recognized and investigated this differential impact of power on women's bodies, as well as how the fantasy of a neutral, unsexed and universal body – which somehow always turns out to be male – conceals that difference. We might conceptualize this secret knowledge concealed within a rhetoric of universality as being *encoded* – that is, the ostensible meaning of something turns out to conceal another meaning, but only after we know how to read it. *Orphan Black*, particularly in its first three seasons, makes extensive use of codes, secret signs and riddles.[2] From the start, there is a riddle, something like a nursery rhyme, that the members of Clone Club use to identify themselves ('Just one, I'm a few. No family, too. Who am I?'), but enigmatic signs multiply – Helena's stick figures, the Prolethean fish – and eventually more literal codes emerge as well (the DNA cipher that Cosima and Delphine break, the strangely annotated copy of *The Island of Dr Moreau* that belongs to professor Duncan). What I want to propose in this essay is that *Orphan Black* repeatedly treats the way that the women's bodies are disciplined, surveilled and controlled as a code that must be broken, a message that is hidden in the apparent neutrality of science, the state – and perhaps television, as well.

The Dancing Men

The classic thriller has often coupled the image of the female body at risk, saved at the last moment by the energetic male hero. While Poe had learned how to narrativize decryption in 'The Gold-Bug', it was not until Conan Doyle's Sherlock Holmes that we would find an enduring model for how cryptographic writing might conceal a crime.[3] (While Conan Doyle is not referenced by *Orphan Black* with the same insistence as some other figures, he is both cited and appears in a photograph in 5.2 – a photo that includes Conan Doyle, the fictional character P. T. Westmorland and one of Darwin's Galapagos tortoises – indicating that he is very much on the show's mind). In 'The Adventure of the Dancing Men', an Englishman, Mr Hilton Cubitt, comes to consult Holmes about apparently harmless, childish drawings that have appeared scrawled about his property, drawings that terrify his young American wife, Elsie, who refuses to speak of them – or of her past. The images are simply variations on the most basic stick figure: a circle for a head, a straight line for a torso, lines for legs and arms. The impression of these stick figures when arranged in a linear series is that of a jubilant dancer performing a jig with occasional somersaults, an impression not reduced by the fact that the figure sometimes holds a festive flag in one hand (see Figure 18). Holmes immediately suspects that the images are not a child's prank, but that they represent a letter substitution cipher, in which each specific variation stands in for a letter, and a flag marks the end of a word. Unfortunately, however, the initial sample that Mr Cubitt presents is too brief to solve the cipher, but as samples accumulate, he learns that 'the dancing men' are messages from a Chicago gangster, Abe Slaney, who claims Elsie for his own. Elsie had been 'pledged' to him years earlier, and he has come to collect what belongs to him

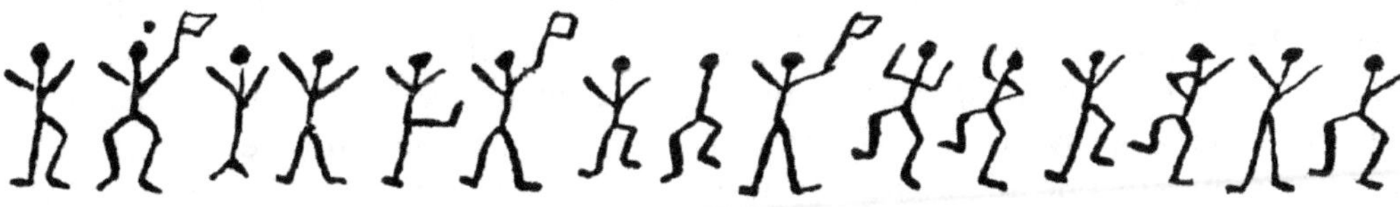

Figure 18: The uncanny 'dancing men' of *Sherlock Holmes*.

(we might note in passing that Abe's name links him to patriarchy [Abraham], and it was Elsie's father who 'gave' her to Abe).

With the aid of a final message, Holmes learns that Slaney has directly threatened Elsie's life, but he is too late to save his client (unusual in the Sherlock Holmes stories). Mr Cubitt has been killed by Slaney, and Elsie is gravely wounded (the reader infers that she attempted to kill herself after the violent confrontation between her husband and her former fiancé).[4] In 'The Dancing Men', masculine writing seems to place woman in the position of the victim in a way that appears to be encoded into the genre – part of its DNA, as it were, written onto and into the body of woman. Indeed, a surprising number of Holmes stories revolve around the question of who gets to 'have' a young woman. 'The Adventure of the Copper Beeches' is about a certain Mr Rucastle who hires a young woman, Miss Violet Hunter, as a governess – provided only that she agree to cut her hair in a certain fashion, wear certain clothes and sit on the sofa of his living room. The strangeness of her duties leads her to consult Holmes, who deduces that Miss Hunter is clearly meant to be mistaken for Mr Rucastle's daughter by the daughter's fiancé, so he will think she has lost interest in him. Mr Rucastle will lose a great deal of money if and when his daughter marries; he demands exclusive possession of her, a possession that is explicitly economic, but implicitly sexual.

The same struggle over possession of the woman, economically and sexually, appears also in 'The Adventure of the Speckled Band'. In that story, a young woman named Helen Stoner consults Holmes about her complex family history. She had lost her father at a very young age, and when she was just two years old, her widowed mother married a certain Dr Roylott in India. Shortly after returning from India, however, Helen's mother died in a train accident, and so Helen has lived most of her life with her stepfather. Since Helen announced her engagement to a young man, a series of strange events (whistles in the night, her step-father moving her to a different bedroom where the bed is bolted to the floor) have made her suspicious and nervous. Holmes deduces that the annual income her mother willed to her is all her stepfather has left, and that he is planning on killing his step-daughter in order to maintain possession of it. The story's uncanny subtext, however, is clearly about his desire to maintain a totalizing possession of his daughter that is both economic and sexual: both the bed bolted to the floor and the means of murder – a deadly snake he keeps in his own bedroom that can sneak into her locked room – speak to his incestuous desire.

The twist that gives this story part of its uncanny charge, and that provides a sense of foreboding, doom or inevitability, is that it has all happened before. Helen in fact had a twin sister (and here we should recall that Leda in Greek mythology gave birth to two sets of twins – including Castor and Helen – and that this is also the source of Helena's name in *Orphan Black*) named Julia, who two years earlier had also met a young man, become engaged, heard the whistling in the night and died a horrible but unexplained death. The uncanny effect produced by the similarity of twins is intensified here by the sense that, to their stepfather, who has a direct financial interest in them, they are as exchangeable as any commodities that he might possess. Helen describes the way in which their rooms are arranged, and it becomes clear that their roles are structurally determined: along a single hallway, there are three rooms, of which the first belongs to the stepfather the second to Julia, the third to Helen.[5] Upon Helen's engagement, her father moves her into Julia's former room, a kind of fatal chamber that marks her as doomed.

Both the uncanny and the question of gender demand that we return to the 'Dancing Men', however. If 'Speckled Band' finds its uncanny in the strange, fated repetition of female twins and in its unspoken desire for totalizing sexual possession of women's bodies and their fates, 'The Dancing Men' finds its uncanny charge elsewhere. Many readers of Conan Doyle's 'The Dancing Men' report that they read this story when they were young and that what they found creepy was not the content, but the *dancing men themselves* – i.e. the images of the stick figures. These apparently blithe and careless figurines represent the most elementary features of the human body – they are childish, but precisely in the sense of betraying no recognition of the awareness of sexual difference. Indeed, initially in the story they are referred to as 'figures' and 'hieroglyphics'. And yet, all the same, Holmes names them 'men'. As we know from bathroom signs around the world, stick figures are male, unless they are marked otherwise by a skirt or hair, yet another way in which the subordinate fate (the female destiny of being possessed by men) is encoded within the code – not in its content, but on its surface, in the very figures themselves.[6] The neutral, universal sign for human is male. Everything about 'The Dancing Men', however, works to conceal both the invisible masculinity of the universal human figure; moreover, the fact that the secret messages are about the male appropriation of the female body appears accidental, rather than part of the code of (masculine) universality.

Codes and Ciphers

Orphan Black amounts to a 'nature vs. nurture' experiment, and the show answers that opposition by very resolutely weighing in on the side of nurture. Over the course of the series, we see what happens to apparently identical DNA when it is raised in a different environment, as we meet more and more of Sarah's clones: Katja, a wealthy German with artsy cosmopolitan tastes; Alison, an uptight and determined Canadian housewife; Cosima, an easy-going, dreadlocked, pot-smoking, bisexual biology grad student from San Francisco; Helena, a psychotic Ukrainian assassin; Rachel, an icy, upper-crust, English CEO-type; Krystal, a smarter-than-she-looks blonde manicurist and even Tony, a transgender drifter. All are, in a purely genetic sense, the same person, and acted by one performer, but Tatiana Maslany's masterful performances produce precisely the opposite effect: *Orphan Black* feels like a large ensemble show.

Not all is well in the nurture world, however. Here, too, female victimhood is written directly into the genes – quite literally. In Season 2, we meet yet another clone, Jennifer Fitzsimmons, through a video blog that chronicles her progressive respiratory illness and eventual death – an illness that also affects the German Katja, the American Cosima and potentially all of the other clones as well. All of them – with the sole exception of Sarah and Helena – are also infertile, and these dual tragedies of illness and infertility (both pre-eminent and long-standing biopolitical concerns about the female body) are both caused by yet another male-created code. The scientist Ethan Duncan, who created the clones in the early 1980s, wanted to ensure that his experiments could not reproduce, and so altered their genetic code accordingly; the progressive respiratory illness is an unintended by-product of that initial alteration. It would appear that biology is destiny after all – at least within the genetic code of the thriller genre, where femaleness is still, on some level, equivalent to victimhood.

In the final episode of the first season (1.10), 'Endless Forms Most Beautiful', Cosima and her lover Delphine analyse Cosima's DNA in a classic code-breaking sequence. I say 'classic' because, just as in Poe and Conan Doyle, the scene moves the viewer from a position of frustration, passivity and helplessness (Cosima learns that she is a mere alphanumeric code to her creators) to one of triumphant active mastery as the code is broken and explained. The two investigators, Cosima and Delphine, want to look at the artificial ('synthetic') sequence of DNA that all of the clones have (the portion that renders them infertile and sick), as well as the small portion of DNA that is unique to each clone and that functions as a kind of 'ID tag':

Cosima:	Okay, here's the synthetic sequence. And this is the differentiated portion.
Delphine:	Yes, this is an encrypted ID tag. It's how they tell you apart.
Cosima:	Okay, but how to decode it? I have four nucleotides to work with: G, C, T, and A. And no idea how they're encrypted. I mean, there's thousands of permutations, and we're running out of time.
Delphine:	Yes, but I know your tag number.
Cosima:	You do?
Delphine:	Yes, I saw it many times. It's numeric, so if we know how that translates, then we can figure out the rest.
Cosima:	Okay. What is it?
Delphine:	It's 324b21.
Cosima:	324b21. Right. I'm 324b21 [*downcast at being reduced to a number*].
Delphine:	I'm sorry.
Cosima:	Well, let's run these letters through a decryption program. [*Later*] Ugh! None of these results match the ID tag. Okay… Okay, think. We're molecular-encoding, from 30 years ago.
Delphine:	Yes, but we're looking at it from now.
Cosima:	Of course! [*realizing the solution*] They weren't coding nucleotides.
Delphine:	It was whole base pairs?
Cosima:	Not *four* letters, but *two*, AT and GC.[7]
Delphine:	Ones and zeros…
Cosima:	Binary. They were coding in ASCII![8] Okay, moment of truth. [*They run the decryption program with the new settings.*]
Delphine:	It's your ID tag. You cracked it!

Code-breaking in detective fiction is generally an unfolding, narrative process, not so much the push of a button and revelation of a message. Typically, we move through a series of progressive decodings, and most often, the final message is something that retroactively restructures the narrative – something that provides a narrative twist rather than resolving the narrative. (This is the case already in 'The Adventure of the Dancing Men'.) At first blush, of course, Cosima's success is figured as a moment of self-realization, a self-appropriation: she has broken the apparently neutral code and literally found herself, if only in the depersonalized form of an alphanumeric ID tag. (On the other hand, one's body – if not one's self – is nothing more or less than the continual process of decoding one's DNA.)

Woman as passive victim is deeply encoded already in the genes of detective fiction and the thriller, however. The men who write the codes come back to take what – in their fantasy, in any event – is theirs. And so, when Cosima and Delphine decode the entire sequence, the message is a message of appropriation cast in the language of the law of the father, which is to say, the legal language of corporations: 'this organism and derivative genetic material is restricted intellectual property'. Cosima and all of her fellow clones have been patented. She explains this appropriation over the phone to her clone sister, Sarah: 'We're property. Our bodies, our biology, everything we are, everything we become, belongs to them'. Sarah declines to accept Dyad's offer of protection and non-invasive monitoring after learning this, and vows to keep herself and her daughter as far from them as possible. Lest the political valence of this sequence be lost on viewers, it returns in 5.5. While sequencing a DNA sample, Cosima has a flashback and we see the sequence repeated and continued. In response to Cosima's despair at being patented, Delphine enjoins her to 'defy them… they will never own you'. If this language of *resistance* sounds familiar in the current political climate, that is deliberate: Maslany clarifies in the short 'closer look' extra accompanying the episode that 'the focus of that episode was… patriarchy and… how you fight it' before going on to note that the episode was made the week after Donald Trump won the US presidential election and that the episode felt like their response to his election (2017d). But the message of the show is not simply one of resistance: it is also a suggestion that the apparently neutral codes of science, genetics and the law conceal a message that is often reducible to 'your body belongs to us'.

The second season of *Orphan Black* amplifies this critique: all female bodies may be appropriated. The 'Proletheans', for example, are a radical religious sect that denies that the clone sisters have souls. One of the sect's leaders, named Tomas, kidnaps and trains Helena to track down her sisters and exterminate them. Another branch of the Proletheans first removes Helena's eggs from her ovaries, and then later impregnates Helena with those same eggs, now fertilized by the leader's semen, without her consent or knowledge. The Dyad Institute takes bone marrow from Sarah's daughter, Kira, to treat Cosima. All of the clone sisters are monitored for years by their romantic and sexual partners who work for Dyad, which probes them and administers medical tests to them in the middle of the night. In short, the narrative thrill of the television show is about finding out and exposing the manifold ways in which women's bodies and reproductivity have increasingly become subject to corporate appropriation and intervention, just as women's bodies are subject to corporate monitoring, generally without their consent or knowledge. (In 2012, for instance, the story broke widely that Target had monitored its female customers' shopping patterns, and on the basis of their purchases alone was able to identify – sometimes before the women told anyone else – when they had become pregnant, which in turn allowed Target to market them coupons for diapers and other baby products.)

Orphan Black has done precisely what Todd McGowan has suggested is necessary in order to mobilize an effective political critique from within popular culture. Rather than attempting to expose the injustice didactically and moralistically with, say, a documentary about corporate surveillance, the show cathects the viewer to the question of the monitoring of female bodies as something glamorous and thrilling ('fascinating', in McGowan's terms), an essential part of the genre (2007: 7–10). In other words, what appears as a thrilling and chilling conspiracy within the show is effectively modelling the viewer's real world, provoking an awareness of social issues without didacticism, relying on the codes of a global genre (and even capitalist entrepreneurship).

Fascination with unravelling the mystery emerges from the basic, tantalizing procedure of detective fiction: there is a mystery, an enigma, a code, and we are only introduced to its solution in fragments. The solution is always deferred, and every time we reach out for it, it retreats one step farther away. In the second season of *Orphan Black*, for instance (2.9), Cosima falls ill with the progressive (and eventually fatal) respiratory illness that is an unintended consequence of the clones' built-in infertility. A series of stopgap measures keep her alive, but only with the help of Ethan Duncan, a scientist who still survives from the original military project, Project Leda, that gave rise to Sarah and all of the other clones in 1984. (We may note in passing that the myth of Leda is yet another story of a patriarch who appropriates the female body to implant fertilized eggs in her.) Ethan has the entire 'infertility sequence' on obsolete 5-inch floppy disks, but even when the clones and their helpers manage to read the out-of-date disks, gibberish emerges.

Cosima: That's the infertility sequence? God, no wonder we're all screwed up! [*Looks at Ethan and smiles to take the sting out of her comment.*]

Ethan: It's encoded. It's a Vigenère's cipher, and I brought the key with me [*taps his pen against his head, indicating where the key is stored*]. I'll need you to generate a transcription algorithm, Scott.

Scott: [*Excited*] This is just killer! [*They laugh and smile at his enthusiasm, but Cosima looks away, and tries to conceal a cough.*]

The Vigenère cipher was regarded for centuries as an unbreakable code. Most fictionalized accounts of code-breaking tend to rely on the most primitive kind of code, the alphabetic substitution cipher, also known as the Caesar cipher, precisely because this kind of decryption can be easily narrativized and explained to the reader. In this form, every letter of the alphabet is substituted by some other letter, usually a fixed 'shift' away (*a* is substituted by *d*, *b* by *e*, *c* by *f*, and so on). Alternatively, as in 'The Dancing Men', letters might be substituted by arbitrary symbols, or by a set of symbols like 1s and 0s. (The genetic sequence that Cosima and Delphine break is a substitution cipher, as is the ASCII standard.) In either case, relatively little exposition is necessary to explain to the reader the basic techniques for breaking a substitution cipher. In Doyle's story, it takes just a few paragraphs for Holmes to explain to Watson the basic rules of letter frequency (the most common symbol will usually represent the letter 'e', for example) and word frequency (a frequent three letter word with *e* at the end is likely to be 'the'), following Poe's explanation in 'The Gold-Bug'. The Vigenère cipher, on the other hand, uses a keyword to generate a new and different substitution cipher for each letter of the text to be encrypted, which tends to eliminate the tell-tale frequencies of common English letters. The exact mechanism of encryption (let alone the method of decryption), however, is too complicated to describe in a television show (or this chapter).[9]

Orphan Black wisely does not attempt to explain the technicalities of the cipher at all. Rather, Duncan equates the code – which stores information about the clone sisters' bodies, their fertility and their health – with himself. 'It's a Vigenère's cipher, and I brought the key with me', he says, tapping his pen against his temple to indicate where the key is stored (men are minds that produce codes, women are bodies that are thus encoded). And indeed, after Ethan begins decrypting the text, he reveals that there are multiple keys, which he will dole out as necessary, and only if Dyad meets his expectations. This endless deferral of the key continues even after Duncan commits suicide, when it becomes clear that the Dyad Institute's only interest is in perpetuating the experiment and creating ever more clones. Before dying, Duncan gives a copy of *The Island of Dr Moreau* to Sarah's daughter Kira (Figure 19). This book functions as a truly unbreakable code, a 'one-time pad', since only this particular copy, heavily and enigmatically annotated and decorated by Ethan, is capable of decrypting the information needed to save Cosima – a task deferred to the third season of the show where, once again, every layer of encoding always conceals another code (indeed, as Figure 19 makes clear, there are multiple codes at work in Duncan's copy of the book).[10] The show's critical stance towards the appropriation of women's bodies is clear – but does it run the risk of a hopeless fatalism? After all, each time the women crack a code, it only reveals another male-written code that claims ownership of the women and their bodies. Moreover, this claiming of ownership never seems to be mere appropriation. Dyad, for example, claims that they will offer a benignly paternal and neutral protection and monitoring of the clones, but in fact, when Sarah surrenders to them, she discovers that this apparent neutrality conceals a hostile invasion of her privacy (doctors demand every detail of her sexual history), and eventually of her body: an operation to remove one of her ovaries for which her consent is neither required nor desired ('enjoy your oophorectomy', Rachel hisses). In the case of Ethan's Vigenère cipher, it is Scott who unwittingly provides the secret meaning to the apparently benign, even avuncular, figure of Ethan Duncan and his code. 'This is killer!' he exclaims with a grin, excited to be in a world of government conspiracies, industrial espionage and human cloning, building transcription algorithms to break secret codes that hold vital secrets – and who wouldn't be? But his exclamation of enthusiasm – 'this is killer!' – is given a very different meaning by Cosima's fatigued body, the cough that

Figure 19: Duncan's secret code book.

indicates the apparently inexorable progress of the fatal disease that this charming father figure has encoded into every cell in her body.

As the show goes on, Sarah and her kin also learn to make use of secret communications of their own; indeed, it is Kira, who studies Ethan Duncan's book quite carefully, who relays a series of coded messages to Sarah at key points through apparently innocent, childhood drawings. The first of these is the image she draws under Cosima's supervision in 2.10, an image that Rachel gives to Sarah to taunt her, a picture of Rachel and Kira holding hands, with a fire extinguisher inexplicably floating nearby (this is the extinguisher that will propel a pencil into Rachel's eye, as discussed by Goulet and Loofbourow in this volume). This kind of secret writing is called 'steganography' – writing that is 'covered' in some fashion so that the code is not recognizable as a code in the first place (see Singh 1999: 6–7). We should recall that, in Conan Doyle's 'The Dancing Men', Watson initially dismisses that code precisely as 'a child's drawing' (Doyle 2005: 865), and this language of the apparently innocent child is sustained throughout the story ('childish prank' [868], 'child's scrawl' [896]). Indeed, by *Orphan Black*'s final season, Kira is able to incorporate all of the codes the show has deployed, from this early instance of steganography to the hidden meanings in childish nursery rhymes. In 5.7, she shows her mother another 'innocent drawing' over Skype, a children's story she is writing about a mouse and an elephant. The animals represent herself and Rachel, as in her earlier drawing, although Rachel does not hold her hand; instead Elephant stands on Mouse's tail, to prevent her from leaving. The accompanying poem – 'Mouse cried and asked to return to the sand / But Elephant said she belonged in

Wonderland' – is a code that informs Sarah that Rachel plans to take Kira to Westmorland's island for further experiments (which in turn are directed, once again, at his attempt to appropriate women's reproductive capacity, in this case, by harvesting Kira's eggs). Once again, the 'childish' drawing conceals not simply a sophisticated meaning, but an 'adult' one: a meaning that is very much about sexual difference, desire and power.

The Dancing Women

When we first encounter the clone Helena in the show, her goals are essentially negative or privative in character: she lives only in order to kill the other clones and cuts herself with a razor blade in an elaborate scarification ritual to give herself angel wings on her back. This negativity comes with its own kind of enigma. After each of her assassinations, Helena leaves behind dolls' heads carefully crafted to look like her victims, as well as Bibles with elaborate circled passages. Her apartment, when discovered, has more Bible verses written in dripping red letters on the wall. So far, this is part of the by-now standard way that psychotic serial killers are represented in film and television, as an enigmatic riddle or code that the investigator must manage to break. (The fact that the killer is female does not change the essential gendering of the code, since the woman here is a vehicle for the male religious fanatic behind her, Tomas, and for the Proletheans more generally.) But, if the Bible verses belong to Tomas, Helena appears to have created an enigmatic code of her own, one which should be familiar to us by now, and it is one that appears everywhere she has been in the show's first season.

Helena's female stick figures are a clear recollection of Holmes's 'dancing men', and like Holmes's figurines, they appear to be an innocent 'childish prank' (Helena leaves them on a handmade toy that she gives to a young boy) and simultaneously contain a message and warning prefiguring an adult female victim. Unlike the figures in Conan Doyle, however, Helena's figures carry an extra mark that indicates their non-universality, the skirt (see Figures 20 and 21). But we may note other differences as well, two in particular. First, the figures do not 'move' or 'dance': they are essentially the same, with no variation of the shape or stance of the arms (thrown out straight to the sides) or legs. Second, they are essentially nonlinear in character. Rather than lines of text, these figures appear to be scattered more or less at random across the visual field, with variations in size. They are perhaps not to be understood as writing at all, then, let alone as a code. Yet they still have a meaning – when Detective de Angelis attempts to parse the hidden meaning of the female stick figures, she sees female victimhood, suggesting that perhaps they indicate 'early childhood development issues' or 'extreme abuse'. Later, she hypothesizes that the figures indicate 'her single victim, our Jane Doe, over and over again' – and, of course, she is correct about both. Helena *does* have early childhood development issues, and she was indeed subjected to extreme abuse; the stick figures *do* represent a victim (we see that some have been crossed out), with the caveat that they are not *precisely* the same victim again and again, but rather a series of clones.

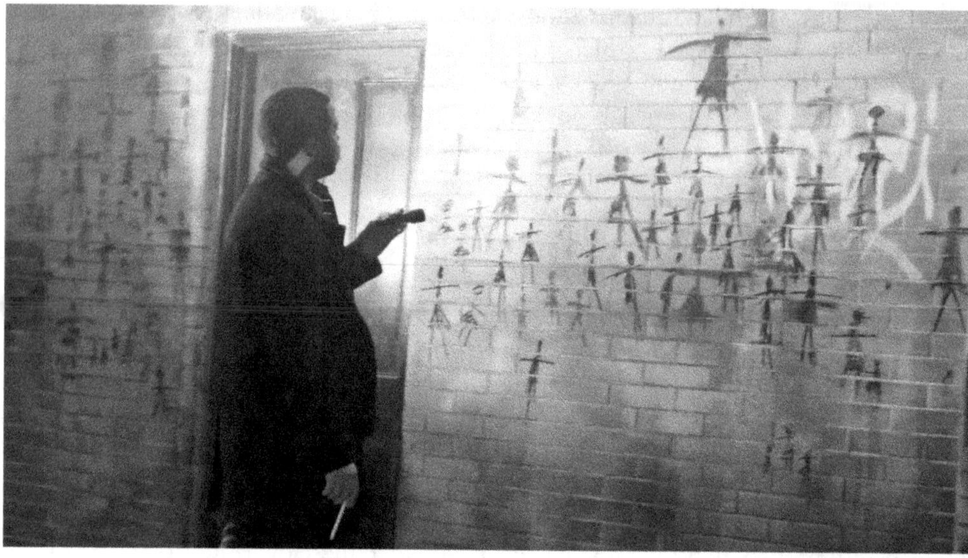

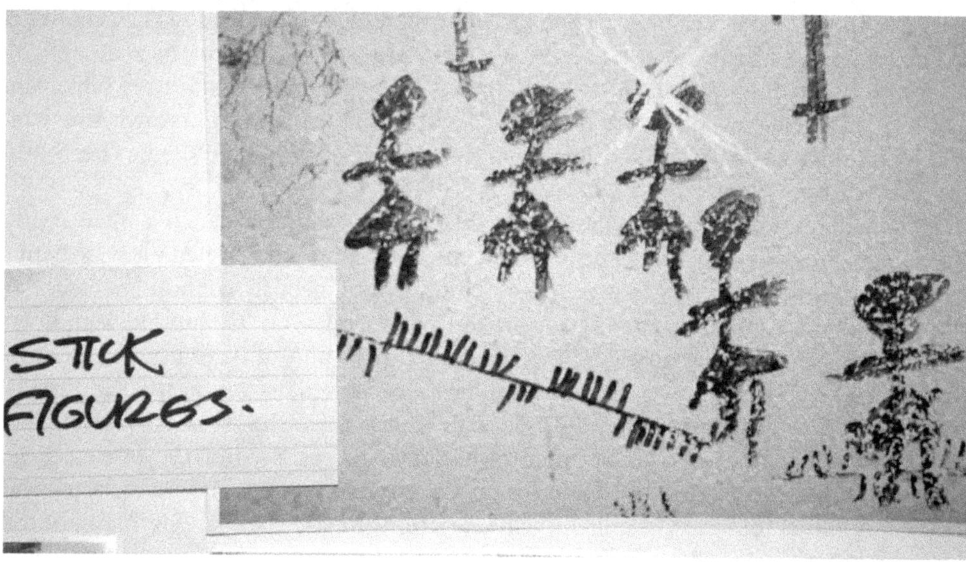

Figures 20 and 21: The proliferation of stick figures, and Katja Obinger's 'cancelled' head.

To be one of these twinned young women would appear to be marked in advance for death (see, for instance, Figure 21, where one figure's head has been crossed out – presumably Katja Obinger, whom Helena shot in the head).

Could there be a way for these skirted stick figures to learn how to become 'dancing women', to have a message that is different from the message of the dancing men, and if so, what would they say? If the dancing men celebrate the way that codes and cryptography have already encoded them as the universal norm, always delivering the same message about the appropriation of women's bodies, could 'the dancing women' find some other message to deliver? Thanks to *Orphan Black*, I believe the answer to this last question is yes. Helena's stick figures are static and unvarying; woman's body is always the same, always to be found in – quite literally for her stick figures – the same position, that of the victim, flat on her back, arms thrown wide. *Orphan Black* is clearly opposed to this conception of woman, however – opposed to the idea that a woman's biology is destiny, that the same DNA code (the clone sisters) always produces the same meaning. Instead, it produces radical difference that in turn produces, in Darwin's famous phrase, 'endless forms most beautiful'. But it does so at the level of the embodied person, not the abstract cipher: Alison, Beth, Cosima and so on are the ABCs that can and do represent a truly Darwinian future – an open one – including figures that break out of this alphabetical sequence altogether (Helena), appropriating a male name (Manning) when it suits them, or even a more male body (Tony). One set of DNA, one body, can in turn give rise to neurotic housewives and stoner biologists, punk rockers with a penchant for crime and buttoned-up police officers, ice queens who yearn to be mothers, and transgender drifters. This production of difference does not occur at the level of (masculine) encoded text, whether it is Ethan Duncan's genetic markers, the Dyad Institute's concealed patents or Abe Slaney's demands and threats – instead, it happens at the level of the body of the characters.

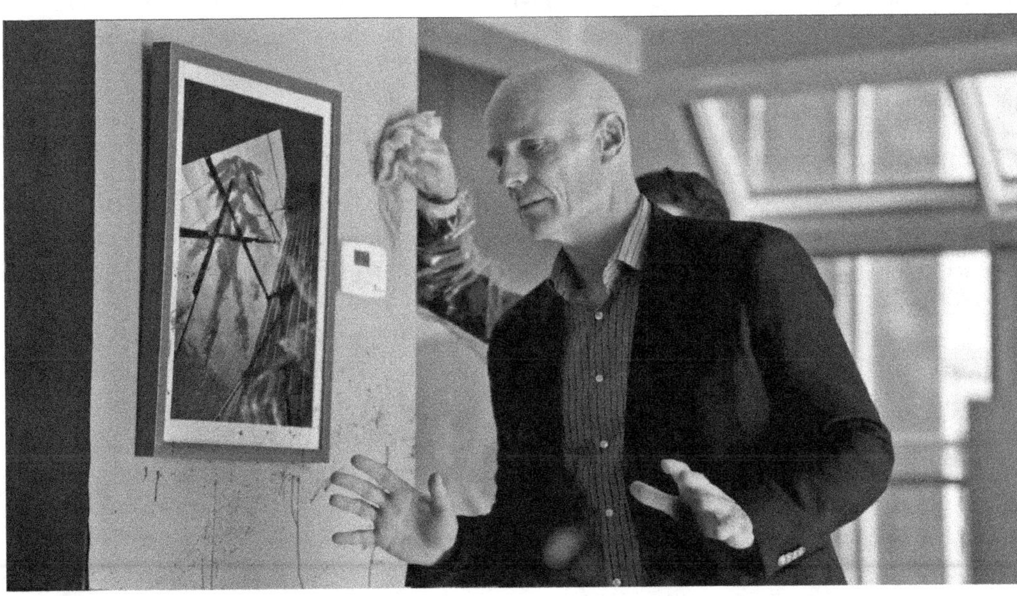

Figure 22: A new interiority.

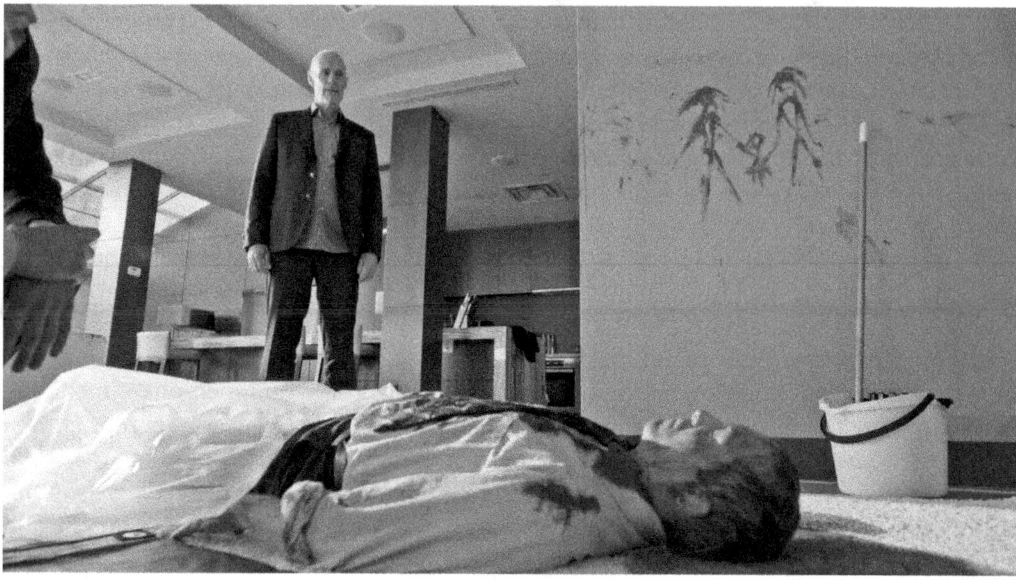

Figure 23: A new family.

Helena learns to stop killing her clone sisters, and when she does, she also stops drawing her blank, identical and endlessly repeating female stick figures. In 2.4 ('Governed, as It Were, by Chance'), Daniel – Rachel's right-hand man – captures and tortures Sarah in Rachel's apartment. As he is about to cut off her ear, the radio inexplicably turns on in a different room, and Daniel goes to investigate. He is killed in a magnificent fight that takes place entirely off-screen and is only heard by the viewer; Helena, still wearing her Prolethean wedding dress, has come to protect her twin *sestra*. In the following episode, we discover that Helena has left stick figures drawn in Daniel's blood all over Rachel's apartment. But they are not at all the same as the ones she used to draw. They are now clearly delineated individuals: Helena is particularly recognizable from the shock of frizzy hair that surrounds her head. The figures have acquired an interior space, particularly Helena, who gives a small but visible 'baby bump' to her figurine (see Figure 22, where we see Helena's assertive self-portrait drawn over one of Rachel's artsy photographs, in implicit contrast to the defensive posture taken by Dr Leekie). And, even more important, they appear in groups now: Helena draws a triptych of herself and Sarah, each holding one of Kira's hands (see Figure 23, with Helena on the left, with her 'filled in' waist and bigger hair, Sarah on the right; this all-female family of three is implicitly contrasted with the three men present, suggesting that the role of victim may not always belong to the woman). The message of the coded figures seems to be quite different than the one in 'The Adventure of the Dancing Men' – not a story about men appropriating women and their bodies, but rather a story about becoming an individual, acquiring interiority, and bonding with others as family. Or perhaps one might argue that this is no longer a code at all: the apparently neutral universality of 'human figures' is gone entirely, in favour of concrete and embodied individuals.

Helena's stick figures put in occasional appearances throughout the rest of the show: they appear dramatically in 3.6 ('Certain Agony of the Battlefield') during Sarah's extended near-death hallucination: they replace what is normally Kira's artwork on Mrs S's refrigerator. In that same hallucination, we see a series of pictures by Helena that show her and Sarah once again holding Kira's hands, an image that appears to be Helena holding her own future daughter's hand, and a portrait of Helena and four of her *sestras* (it is not clear who the fifth clone is here, but possibly Rachel). These are evidently products of Sarah's imagination, but we also see in 3.5 ('Scarred by Many Past Frustrations') that Helena, believing Sarah had betrayed her to the Castors, has decorated the inside of her cell with another drawing of herself and Sarah – this time, with Sarah crossed out. They never go away – in Season 5, we see that Helena has decorated her cell in the nunnery where she takes refuge with more of these images: another group portrait of the clone *sestras*, herself with her future twins, what appears to be herself escaping a cage, a portrait that might be Helena with her boyfriend Jesse and so on. Indeed, the stick figures bookend Helena's chronology within the show: in 5.9, we see the figures in her 'origin story', drawn on the basement door where she is first held prisoner by the nuns (the image is a solitary figure next to a church), while in 5.10, they return in angelic form in a mobile that hangs above her babies' cribs. In each case, these images function as indices of Helena's cognitive and emotional development, but they also constitute a continuing refusal of the universalism present in Conan Doyle's 'dancing men'.

In many ways, however, the dialogue between Helena's images and Conan Doyle's code reaches its culmination in the second season's finale. In that episode, Helena is finally brought to Felix's apartment to meet all of her principal clone sisters at once – Alison, Cosima and Sarah. They are united in working together to save Cosima, protect Kira and undo the 'unintended consequences' of their own genetic code. Over the course of the first two seasons, they have slowly brought onto their side some of those who initially worked to maintain control of their bodies and their destinies (Donnie, Alison's husband, or Delphine, Cosima's girlfriend). Things are not great – Cosima wears an oxygen mask and has trouble standing, powerful and shadowy agents seem to be united against them, and the police suspect their involvement in several murders. Even so, Cosima puts on a record, removes her oxygen tubes and turns to face her sisters. Everyone watches as she sways to the beat, and Felix comes up to take Cosima by the hand and draw her onto the makeshift dance floor. And in the recognition of their solidarity, their common interest in owning and controlling their own bodies, their own genetic codes, their children and themselves, they all rise up together and – in one of the most celebrated (and rightly so) sequences of recent television – they begin to dance, each in his or her own way, and all together. There is a message here in these dancing women, and it is not hidden: an affirmation of both difference and solidarity – and a celebration of the embodied person, rather than the abstract figure.

Biopolitical Fantasy

What, then, is the biopolitical fantasy about codes and decoding, about bodies, health and vitality, offered by *Orphan Black*? On the one hand, the term 'fantasy' suggests a kind of wish fulfilment. *Orphan Black* offers multiple examples of this sort of biopolitical fantasy, including Dr Leekie's vision of enhanced 'post-human' humanity that can see in infrared and add whatever body modifications seem appealing that day; in quite the opposite direction, BrightBorn imagines a future where every baby has dimples and perfect health (and is conspicuously white). In the show, of course, these are revealed as fantasies that are incompatible with reality: the genetic fantasy of BrightBorn, for instance, relies on a very dark underbelly of radically disfigured and disabled infants carried to term by poor women (generally of colour – a perfect example of Mbembe's argument about 'necropolitics': the biopolitical production of affluent western health and vitality is reliant on the concomitant production of disease and death elsewhere). Indeed, philosopher Slavoj Žižek has long insisted that fantasy is more properly understood as a kind of staging of not only what we desire, but why we cannot attain it. In his discussion of the idea of ideological fantasy, he suggests that we always overlook 'the illusion which is structuring our real, effective relationship to reality' (Žižek 1991: 32) – an illusion that always follows the same essential formula: we would be happy if not for X (where X can be variously filled in with feminists, misogyny, affirmative action, Monsanto, abortion, guns, neoliberalism, Donald Trump, etc. – i.e. whatever the phantasmatic barrier to our satisfaction might be). What is remarkable about Žižek's observation, however, is that the phrase 'if not for X' is an indispensable element of the fantasy; indeed, it is what constitutes or allows the fantasy in the first place. In order to even imagine the satisfaction of my desire, I have to first imagine that satisfaction being blocked. The case of biopolitical fantasy is no different: a utopian ideal is held out – the clone community of *sestras*, in which truly all are different, but all are (literally) equal, everyone loves everyone else as they love themselves – but the fantasy lies not in this utopian scene of satisfaction, but rather in the thrilling ways it might be blocked. This is yet another form of the temporal doubling that Väliaho sees in the biopolitical image (2014: 58), which contains implicitly an idealized future that we hope to reach (and that somehow always slips out of grasp), but whose manifest content is a threatening present that demands pre-emptive action.

As I suggested earlier, code-breaking is often an exciting part of the thriller or the detective narrative, but the solution of the code does not normally solve the crime or exhaust the narrative. Indeed, it often – perhaps even normally – leads to the *loss* of satisfaction, from Conan Doyle's 'The Dancing Men' (death of the client, attempted suicide of his wife – indeed, one of Holmes's most celebrated failures) to the dancing women of *Orphan Black* (their collective dance ends in Helena's abduction and a near-death experience for Cosima). After professor Duncan's suicide, the code that Cosima and Delphine have begun to break endlessly retreats out of their grasp throughout Season 3 of the show in a manner typical of a Hitchcockian MacGuffin: an object that is pursued by the main characters in order to advance the plot, but is in fact of little narrative consequence (a concretized version of the fantasy that often simply disappears or is revealed to be meaningless). Duncan's copy of *The Island of Dr Moreau* is encoded with what is supposed to be the key to the infertility sequence, but that code actually leads to a second code, which can only be deciphered by Rachel. That second code, when deciphered, then generates an enigmatic poem ('In Londontown, we all fell down / And Castor woke from slumber / To find the first, the beast, the curse / The original has a number'), whose hidden meaning must be decoded, and which is followed by… an enigmatic alphanumeric code (H46239). This eventually leads the *sestras* not to the infertility sequence but to Kendall Malone, the original source of both Castor and Leda DNA, who serves as the MacGuffin of Season 4 before she, too, is definitively removed, and another new cure is dangled as a possible source of health, life, vitality, fertility – a cure that will once again move out of reach by the end of the season… and so on.

Indeed, perhaps the show's ultimate biopolitical fantasy is simply that of a normal life (whatever that would mean – presumably something quite different for each of the *sestras*). Again and again, Sarah declares that she 'just wants out of this shit', usually right before she is reluctantly dragged back into it. On the one hand, 'this shit' is exactly what the viewer comes for – the more extraordinary chases and desperate escapes, and the more extraordinary freaky cybernetic worms implanted in the face and men with living tails, the better. On the other hand, what I have been suggesting throughout is that the extraordinary events and characters of *Orphan Black* are in fact disguised (we might even call them encoded) ways of talking about just that: ordinary life. While few viewers are pursued by multinational corporations and religious sects, those groups do have a vested interested in biotechnology, and particularly in female reproductivity. While few viewers are monitored in secret by their partners in the middle of the night, it is certainly true that we live in an unparalleled moment of surveillance, a surveillance that has a very different impact on women and their bodies. The idea that images of your body might be viewed by strangers over digital networks (or the even more invasive possibility explored in 5.7 in which our own eyes become instruments for other people's voyeurism), for example, is one that affects men and women quite differently. And while few viewers will be called upon to break secret codes that are literally implanted into their DNA, I have tried here to suggest that *Orphan Black*'s playful reformulation of Conan Doyle's dancing men conceals a secret message that is both incredibly important and yet so banal that its universal character makes it completely invisible – the seemingly universal abstract figure masking an ill-concealed gender, and always the same one. If *Orphan Black*'s biopolitical fantasy is a return to normal life, the show's political lesson is about just how problematic that 'normal life' can be.

Notes

I'd like to acknowledge my fellow panellists and attendees at our *Orphan Black* panel at the 2016 ACLA conference for their helpful comments and insights, as well as two sets of students: the graduate students in my 'Biopolitics and the moving image' seminar, and the undergraduates in my 'Persons and things' seminar for the Campus Honors Program – both made exceptional connections and observations, and posed insightful questions about *Orphan Black.*

1 A number of science-fiction television series in recent years have dramatized precisely this doubled biopolitical temporality through time travel: the present becomes supercharged with biopolitical urgency in series as disparate as the first season of *Heroes* (2006–10), and more recent shows such as *Continuum* (2012–15) and *Travelers* (2016–). In the more recent versions, travellers come from the future to the present with the express intention of forestalling the biopolitical catastrophe of their time, often something that threatens the survival of the human species. In cinema, of course, this sort of biopolitical image emerges from Chris Marker's *La Jetée* (1962), and continues through the various *Terminator* films (1984, 1991, 2003, 2009 and 2015, to date). Television's serial format, however, has tended to exasperate the premise; in *Travelers*, for instance, the team labours constantly to fulfil their mission, but the future always appears irreparable, beyond their reach.

2 There is a technical distinction in cryptography between *codes*, which replace whole words with other words (arbitrarily chosen), and which require a dictionary to crack, and *ciphers*, which use an algorithm to substitute every letter of a message with something else (another letter or symbol) (see Singh 1999: 30–31, on types of hidden writing). I do not use the term 'code' in its technical sense here, but more to refer to conventions that are implicit and not stated openly ('the codes of detective fiction'); I do, however, use the term 'cipher' in its technical sense throughout. Morse code is, according to this strict definition, not a code at all, but a typical example of a substitution cipher.

3 Rosenheim argues that Poe's interest in and writings on cryptography 'form a matrix out of which not only the language, but the plots and much of the generic form of the detective story emerged' (1989: 376), suggesting that Poe's thinking about code-breaking informs the genre's structures at a very deep (indeed, hidden or encoded) level. Rosenheim concludes his essay with a letter from a correspondent that Poe included in his final note on secret writing, which notes the value of secret messages to 'the statesman and the general – to the scholar and the traveller – and, may I not add "last though not least", to the *lover?* What can be so delightful… as a secret intercourse… a *billet doux* that will not betray its mission, even if intercepted…?' (quoted in Rosenheim 1989: 392). We might, however, take Poe's correspondent's words seriously and consider that the breaking of the cipher or code in detective fiction is an essentially *sexual* scenario that trades on the idea of a resistance that will 'open up' a secret pleasure once overcome.

4 Tellingly, Rosenheim's account of 'The Dancing Men' does not mention Holmes's invocation of gender in the solution to the cipher, or the fate of Elsie and her husband (1997: 54–56).

5 Irwin notes that 'The Speckled Band' itself is an uncanny repetition – not only does it repeat the scenario of the locked room with (at least potentially) two female victims from 'Murders in the Rue Morgue', substituting one animal for another, but the presence of the stepfather indicates 'just enough Oedipal resonance' to show that Conan Doyle 'had some inkling of what was at issue' in Poe's story (1994: 431). The reading I offer here of 'The Adventure of the Speckled Band' suggests that Conan Doyle was not simply echoing Poe in an inferior

way (which is how Irwin and Rosenheim generally treat him), but doing something equally uncanny and not exclusively Oedipal.

6 It is impossible to think about this issue – the elementary symbols that denote male and female – without recalling the notoriously difficult French psychoanalyst Jacques Lacan's discussion of this issue in 'The instance of the letter in the unconscious'. There, Lacan gives the humorous story of a train car with a little boy and a little girl that pulls in at a station (2006: 416–17). The children mistake the signs for the restrooms on the platform as the sign indicating the place they have arrived in, so the boy believes they have arrived in the town of 'Gentlemen', and the girl, the town of 'Ladies'. Lacan uses this anecdote to demonstrate the arbitrary nature of the linguistic sign (we could flip the words, call men 'ladies' and women 'gentlemen' and the world wouldn't change), but also the (false) sense of inevitability and 'rightness' that accompanies sexual identity – the arrival in the magical town of 'Gentlemen' or 'Ladies' – in a heteronormative world. Lacan's point is that, although the symbol on the doors may vary, the doors themselves are the same; switching the placards changes nothing.

7 In fact, this is precisely how DNA works: although there are four 'letters' (A for adenine, C for cytosine, G for guanine and T for thymine), they form two chemical pairs across the double helix, so that A is always paired on the other strand of the helix with T, and C is always paired with G. Such pairs could effectively be treated as binary, allowing any arbitrary message to be written into a DNA sequence.

8 ASCII stands for 'American Standard Code for Information Exchange', which is how computers encoded letters and numbers in the 1980s. ASCII is effectively a substitution cipher: each letter in the alphabet is assigned an essentially random sequence of 7 characters, 1s and 0s only, so that the lower-case letter A might be 1100001, while 1001100 would be an upper-case L. Beginning in the 1990s, a different standard began to emerge for alphanumeric encoding, known as Unicode. ASCII is still in widespread, if waning, use today because it only allows for a relatively small number of characters (128, which suffices for lower and upper-case letters, numbers and punctuation), while languages like Chinese require many more.

9 See Singh (1999: 45–99), for an explanation and the cipher's history.

10 For an explanation of a one-time pad, see Singh (1999: 120–24, although much of the preceding chapter will be necessary to understand the explanation). For an extended (and quite fascinating) account of the use and practical development of one-time pads in the Second World War, see Marks (1998).

The Replicant's '*Réplique*': Motherhood and the Posthuman Family as Resistance in *Orphan Black*

Andrea Goulet

Gynaecological stirrups loom large in *Orphan Black*. In the show's first three seasons alone, seven separate scenes feature a subdued body with legs raised and held in cold clinical clasps while a medicalized phallic object is prepared for forcible insertion. Cosima's stem-cell treatment, for example (2.7), consists of a displaced lesbian act of penetration by Delphine, who lovingly caresses and kisses her as a female doctor guides a lubricated wand containing bone marrow into her uterus. Two episodes later, the pubescent Gracie Johannsen finds herself in stirrups in a barn-turned-hospital room, with her cult-leader father Henrik in the unsettling position of inserter of a fertilizing wand; when Henrik subjects the captive Helena to a similar operation, it is to extract genetic material for external fertilization, in preparation for reinsertion, again by forcible means. And, of course, who can forget Helena's sadistic revenge, the horror-movie shots of Henrik bound and spread-legged, as his wild-eyed victim-turned-torturer wields a horse-sized syringe in a victorious gender reversal of the violation. As feminist critics have suggested since Luce Irigaray's *Speculum of the Other Woman* (1985), the female anatomy, named as the site of male conquest, can also become a force for re-empowerment through a *détournement* or 'hijacking' of the tools of the master. It is not by reversal but by evasion and displacement that Sarah Manning is able to escape her own horror scene of gynaecology in the dramatic finale to Season 2. Her ordeal at the Dyad Institute, where Dr Nealon plans to remove an ovary to harvest one of her eggs, begins with verbal invasion, through a battery of questions about reproduction: 'Are you ovulating?' 'When did you first have sexual intercourse?' 'Have you ever had an STD?' 'Have you ever been diagnosed with a reproductive disorder?' 'Have you ever had an abortion?' 'Do you use birth control?' By the time she is in restraints on an operating table, Sarah understands that she will be subjected to a forced oophorectomy (the extraction of an ovary – only one, assures Dr Nealon, since 'we don't wish to render you infertile'). But her friends have an ingenious plan that allows Sarah to escape – and her ordeal ends with another dramatic scene of displaced violation, as a sharpened pencil gets launched into the enemy Rachel's eyeball, so that the viewer can enjoy the libidinalized revenge-rape fantasy of a long sharp instrument lodging itself in the soft, humid, egglike substance of an unwilling female body. While enucleation can be read symbolically as a sign of castration in males, in this case the destruction of one eye/'one egg' not only displaces Sarah's oophorectomy onto Rachel, it may well signal (or presage to the viewer) the deep trauma presumed to be motivating Rachel's rage: her own infertility.

Sarah Manning's near-oophorectomy and Henrik Johanssen's breeding experiments are joined in Season 3 by more scenes involving reproductive vulnerability, as we learn of army doctor Virginia Coady's human sterilization weapon in the form of male clones (the 'Castors') whose sexual contact with female victims leaves their ovaries atrophied. And in Season 5, Cosima and Delphine are horrified to learn that P.T. Westmorland, aided by Rachel and Susan Duncan, plans to harvest eggs from the prepubescent child Kira in order to implant them in 1,300 test subject surrogates (5.5). These are all dramatic, often shocking, plot devices that tap into visceral fears – but they also serve a larger philosophical function: through each of its seasons and most of its characters, *Orphan Black* stages the scientific and ethical limit-cases of biological procreation and its suppression.

'Fertile, infertile: they're all just people, love'. Siobhan Sadler [Mrs S], mother-figure to many of *Orphan Black*'s characters, utters this key line to Gracie Johannsen, who has been taught by her Prolethean father to believe that clones are 'abominations' (in part, because they are mostly infertile). The assertion – that fertile or infertile, they are all just people – functions at a number of different levels. Primarily, it argues for the clones' fully human status – and it does so by dismissing a distinction based on their capacity to reproduce biologically. This is worth noting because typical contestations of clone humanity most often point to their *origin*, as scientific products, rather than to their capacity to 'bear fruit' in the future. But given the twisted Biblical ideology that Gracie has inherited from her father, Mrs S's statement also specifically rejects the religious glorification of motherhood – and indeed, of virgin birth, which the Prolethean cult leader Johanssen had used to justify the forced insemination of multiple women (including, as we have seen, Helena and his own daughter), reducing them to the role of 'brood mares'. And in fact, though Mrs S says 'they're all just people', she may as well have said '*we*'re all just people, love', because her statement also serves to reassure Gracie that she herself has not irrevocably fallen from grace by virtue of having suffered a miscarriage. Indeed, the viewer soon learns that Gracie's own infertility, rather than being the sign of divine castigation, was transmitted to her through sexual contact with the male clone Mark who, like his brother-clones, carries a pathogen protein used by the military for human sterilization trials on female victims whose symptoms include vaginal bleeding and atrophied ovaries. Nor, of course, are the Castor boys' sexual partners the only sufferers of uterine disorder; as we know from Cosima's treatment at the Dyad Institute, the female clones of the Leda Project are dying from growths that originate in the womb.

Or at least most of them are. One of the series' key revelations is that Sarah Manning and her biological twin Helena have somehow escaped their makers' enforced sterility. Not only has Sarah given birth to a child, Kira, but Sarah also turns out to be inexplicably immune to the Castor pathogen: when injected with the clone brothers' blood, she suffers a nearly fatal fever but survives with none of the expected symptoms of reproductive disorder. 'Kira is a gift', Mrs S tells Sarah (1.2). And through that 'gift', Sarah becomes a kind of postmodern secular miracle, an ironic subversion of the Virgin Mary trope – because, after all, when *all* conceptions can be immaculate (in the test-tube world of a Dyad or a Topside), a return to good old-fashioned sex as the means of reproduction can in fact constitute its own form of rebellion. From the tongue-in-cheek Virgin Mary statue that Felix puts at Sarah's shrine at her fake wake to Helena's blood-crazed reliance on the Bible verse 'knitted in my mother's womb' (1.3), the purity of traditional biological reproduction is problematized and subverted. Moreover, despite Mrs S's claim that motherhood doesn't matter, the opposition between fertility and infertility actually *does* constitute the key binary at play in the *Orphan Black* text – perhaps not in the theological sense meant by the Proletheans, but certainly as the crux of the show's narrative logic.

Before turning to how that central binary fuels the show's potential as political intervention, let me propose an outline of the structural logic of reproduction in *Orphan Black*. Functionally, the female clones in the show can be arrayed across a spectrum of biological reproductive capacity that might be seen to veer towards aligning motherhood with virtue and sterility with vice. On the 'good' end of the scale, Sarah can conceive – and has done so. Her first act upon exiting the train at the start of 1.1 is to call for the right to see her daughter. And by virtue of her affective link to Kira, Sarah remains the show's emotional centre throughout its three seasons. Indeed, despite a rather Deleuzian dispersal of identity through the replication of clones, Sarah is *Orphan Black*'s heroine, the central node in the show's character, who is respected, loved, desired or feared by all in her orbit – and this most often by virtue of her exceptional role as mother. Alison is a mother, too, although in her case it is through a work-around to deal with her sterility: she has adopted two children who, unlike Kira, are relegated in the show to the status of plot device and comic relief. (They are notably absent, for example, from the dinner party at Bubbles in Season 3's final episode, during which Alison says 'whatever comes next, we'll face it together, as a family' – Sarah's child, on the other hand, is invoked and implicitly looped into the alternative family circle by Helena's toast to 'Kira's little face'.) As for Cosima, biological sterility might be linked medically to the threat to her life, but homosexuality allows her to enter an alternate realm of positive non-procreation. The same might be said of the trans gender clone Tony, who has turned his back on female organs of reproduction without forsaking the pleasures of sex, flirtation and foreplay. Helena's case is trickier: she starts out as a death-dealing monster, but this crazy yet compelling Ukrainian ends up attracting viewer sympathy through her intense bond with Sarah, a bond that spans both ends of the conception/reproduction cycle. As foetuses they shared a womb, lying through the gestation period in the head-to-toe twin position that they would later recreate in a camping tent (2.6); and as adults, they share a baffling fertility and maternal status – think here of Helena's reciprocated tenderness for Kira and also of her wonderful pastel sugar fantasy of a baby shower (3.1), which slyly subverts cloying narratives of maternity while allowing the viewer to see (perhaps even to share) Helena's joy at impending motherhood.

But situated clearly at the opposite end of the spectrum of fertile fulfilment, we find Rachel Duncan – the cold and corporate 'pro-clone' bitch, whose complete lack of human warmth, indicated by steel-and-glass exteriors and sharp cuts of costume, turns out to be compensation for her deepest frustration, that of not being able to conceive. In a key scene (2.8), Rachel confronts her father/maker Ethan Duncan about the gene sequencing code that could heal Cosima and wonders why the 'unmonitored tramp' Sarah has been the only one able successfully to conceive. She is no success, responds Duncan, but in fact a scientific

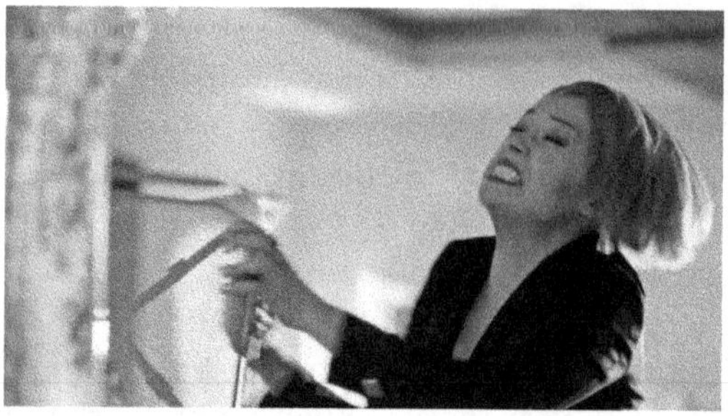

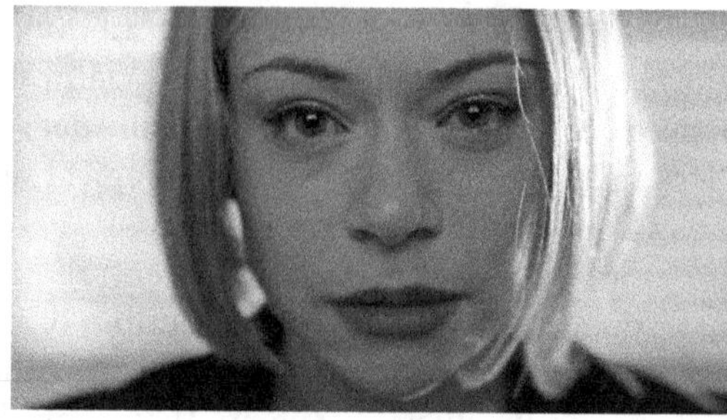

Figures 24 and 25: An enraged Rachel lays waste to Leekie's office, then returns to an eerie calm, just a few hairs out of place.

failure – since the clones are 'barren by design'. This revelation sets off a series of camera cuts between two Rachels, one who continues her steady and calm conversation with Duncan and another who howls with despairing rage, thrashing and trashing her orderly office in a tearful tantrum (Figures 24 and 25). Whether we read the scene as displaying Rachel's inner turmoil versus outer calm or as cutting between the present of her conversation with Duncan and an actual, later outpouring of rage, her repressed but violent emotions upon learning that she is 'barren by design' expose the depth of Rachel's procreative urge, this fundamental biological capacity denied to her by her earthly maker. The enforced sterility of the clones is presented as an act of dehumanization, the brutality of which Duncan does not seem to comprehend; later, when he explains to Cosima the experiment's 'sterility concept' of degrading the endometrium to prevent ovarian follicles from maturing, even the brainy non-breeder Cosima furnishes ethical outrage in the form of her one-word reply: 'Horrifying'.

Reproductive freedom is sacrosanct, then – a stance unsurprising in our twenty-first century age of feminism. But more surprising – even, perhaps, disturbing – is *Orphan Black*'s seemingly conservative reverence for motherhood. (Rachel to Sarah: 'You're very lucky. Motherhood is wonderful' [1.10]; Cosima, admiringly, to Sarah: 'You propagate against all odds' [2.10]; Helena, fiercely defending the canister of her frozen eggs: 'These are my babies' (3.8) and 'you should not threaten babies' [3.9]; Alison's jealousy of Helena's impending motherhood [4.2].) How can the structural spectrum from heroically fertile Sarah to steely, sterile Rachel be reconciled with the progressive sexual politics that the show supposedly promotes? In other words: does *Orphan Black*'s continual return to the question of fertility as an inviolate right undercut its potential for posthuman radicalism? I would propose that the answer to this question is no. And I would go further, by putting forth the thesis that *Orphan Black* actually succeeds in articulating radical alternatives to traditional kinship structures *without* loosening the bonds of (potentially retrograde) maternal ideology.

In order to support that thesis in the remainder of this chapter, I put *Orphan Black* into dialogue with two fictional intertexts (Jean-Michel Truong's 1989 *Reproduction Interdite* and Maurice Dantec's 1999 *Babylon Babies*) and two theoretical frames (Catherine

Malabou's work on biopolitical resistance and Irving Goh's theory of the 'reject'). Unlike the explicit Anglophone intertexts of *Orphan Black* (see this volume's introduction), the French-language sci-fi cloning novels by Truong and Dantec may or may not be known to the show's writers (although they must certainly be aware of *Babylon A.D.*, Mathieu Kassovitz's atrocious 2008 film adaptation of Dantec's novel). But regardless of the question of direct influence, these two early clone fictions serve as suggestive frames that allow us to examine post-biological reproduction – and more specifically, the character dyad of fertile heroine/barren bitch – as a gateway to reconceptualizations of the womb as textual matrix. They also introduce the concept of the *réplique*, or tactical response, which I connect to Malabou's reflection on the links between cloning and 'the reserve of possibilities inscribed in the living being itself' (Malabou 2016: 431). Such possibilities are the crux of Irving Goh's recent MLA prize-winning book *The Reject: Community, Politics, and Religion after the Subject* (2015); through the logic of the 'reject', *Orphan Black*'s Clone Club emerges as a form of countersovereign, deterritorialized 'negotiated resistance' to neoliberal power – a resistance based on new definitions of family, motherhood and (re)generation.

La Réplique: Truong, Dantec and *Orphan Black*

Considered by some the first French posthumanist novel, both for its central theme of industrialized cloning and for its form (which combines corporate memos with Artificial Intelligence [AI] computer-generated dossiers), *La Reproduction Interdite* ('Reproduction not allowed') was written by the Alsatian-Vietnamese novelist Jean-Michel Truong, who was also the founder of the first European AI corporation, 'Cognitech'; the novel was awarded the 1989 Mannesmann-Tally literary prize for novels relating to informatics. Set in the mid-twenty-first century, *La Reproduction Interdite* exposes in detective novel form the nefarious crimes perpetrated under cover of a corporation called 'Reproductique SA' that has established dominance in the European market by fabricating human clones by the millions. The company's official business plan describes the commercialization of clones destined for the health sector through a four-pronged marketing of their product, which includes embryos, foetuses and living or frozen subjects at various stages of post-natal maturation; the medical markets include the pharmaceutical industry and medical research labs, hospitals catering to mass surgery in cases of a bomb or other violent attack, transplant departments wanting to ensure freshness by the direct extraction of organs from living clones, and the deluxe elective surgery sector in which wealthy patrons pay for a cloned 'twin' to be raised in a specialized centre to the age of 40 and cryogenically conserved for harvesting. Truong's futuristic society generally accepts the use of clone skin and other body part by-products for use not only in cosmetic surgery, but also for the fabrication of luxury goods like human leather purses. And Reproductique SA prides itself on the scale of its mass production: nearly a million litres of blood yearly and tens of thousands of adult clones designed to provide hard labour in risky occupations, serve as sacrificial troops in military raids and function as strategic carriers of biological weapons. But investigators in the novel uncover other, less licit, activities as well. In particular, a secret project called 'Plan Fleshware' crosses the line of legality by creating celebrity doubles – that is, clones made to be prostituted to clients wanting to sleep with their screen idols. And at an even more hidden level of depravity, the novel's protagonist discovers the shocking hobby of clone-hunting, in which naked adult human clones who have been conditioned with fight-or-flight reflexes are let loose on a large private estate to be tracked and killed for sport. This cruel application of cloning technology is the brainchild of the cold and powerful Agnès Verges-Ballin, daughter of Reproductique SA's founder – herself a clone, and now the company's CEO. The parallels between Agnès and *Orphan Black*'s Rachel are striking. Like Rachel, Agnès is a clone who enjoys special status because of her upbringing and who lashes out with jealousy at her clone sisters. In her case, however, the sisters number in the tens of thousands; after Agnès's artificial conception, her father extracted a sample from the embryo and when he was later ousted from the company by his daughter, he thawed the sample and brought to term 10,000 Agnes-clones as revenge. (They are all named, jokingly: Agnès-Marie, Agnès-Raymonde, Agnès-Lucie, etc.) Agnès is more than happy to have those clones exterminated in droves – with as much equanimity, perhaps, as Rachel in the 'Helsinki' backstory that is fleshed out in the second series of the *Orphan Black* comics. But more importantly, Agnès is like Rachel, in that she is also 'barren by design'. As explained in its corporate business plan, Reproductique SA abides by current jurisprudence decisions by rendering their copyrighted merchandise sterile: all embryos, they declare, 'satisfy the criterion of reproductive non-autonomy stipulated by the court' (Truong [1989] 2015: 136, all translations mine). Thus the book's title, 'Reproduction not Allowed', takes on a double meaning – of corporate copyright exclusivity as well as of an imposed restriction on the company's human product.

But is the product human? That is the question posed at every turn in Truong's novel, for it is the clones' humanity that is at stake if their mistreatment is to be considered a crime. *Reproduction interdite* provides dozens of pages of (fictional, twenty-first century) theological, judicial, corporate and scientific debates about the human status of clones in which the question of reproduction and reproducibility takes on a crucial role. In a theological discussion reminiscent of Hank and Tomas's debates in *Orphan Black*, one of Truong's characters claims that

> as opposed to animals, [clones] were not placed by the Creator in his Eden nor were they given a spot on Noah's ark among the creatures that God deigned to save from the Deluge. They are nothing but chimaera, born in laboratories, for the use of laboratories. But they are perfect replicas [*répliques*] of mankind.
>
> (Truong 2015: 75)

And in a court trial surrounding the Plan Fleshware prostitution ring, judicial debate pits the prosecution, which wants to grant clones the protected status of animals, against the defence, which claims they are merely non-sentient objects with patent (and patentable) use-value. (Remember Cosima's shock at learning that she and her clone *sestras* are legally owned by the Dyad corporation, organisms derived from patented genetic material.) Here, again, Truong's novel employs the term *réplique*, as the court rules in favour of the defendants because insofar as the prostituted clones have not been proven to be definable as human nor animal, they cannot be considered victims of criminal abuse: 'this activity concerned the use merely of living replicas [*répliques*] of famous men and women, produced by artificial manipulation' (2015: 89). Significantly, it is their lack of reproductive capacity that distinguishes clones not only from living humans and animals, but even from their source ova:

> The Court recognizes the legitimacy of the use of living human organisms as long as it is proven that they have been obtained by the simple replication of a human ovum [...] and as long as it is proven that the organism obtained by this method cannot autonomously reproduce itself.

This court ruling is the reason behind Reproductique SA's 'barren by design' policy: 'for obvious reasons of copyright protection, the embryos are rendered sterile, and therefore satisfy the criterion of non-autonomy of reproduction decreed by the Court' (Truong [1989] 2015: 135–36). Jurisprudence declares that 'the protection of the law accorded to a human ovum is not applicable to its artificial *répliques*. Indeed, these *répliques*, insofar as they are man-made, come under the jurisdiction of common law pertaining to fabricated objects' (Truong [1989] 2015: 136).

I have retained the French word *réplique* in these citations because while the court uses the term to designate a replica or replicant – i.e. a mere copy – the word also contains in French the alternate meaning of *a retort, a response, an objection* to imposed order (visible in the related English word 'reply'). And here is where we can begin to see a potential for subversion in *Reproduction interdite*'s clone population, through the philosophically important shift from object to subject, from material product to human agent. One might argue, in fact, that the court's judicial language carries within itself the seeds of dissolution of its own authority. For a *réplique* is also a revolt – and that shift from object to agent is staged in Truong's novel by a clone mutiny. At the end of *Reproduction interdite*, investigators are shocked to learn that the clones being held in one of the corporation's production plants are not only sentient, but are reacting against their oppression through strange acts of solidarity and resistance. When factory workers come to collect a clone for use in the world, the other clones gather round to try to stop the 'extraction', often at risk to their own lives. A number of clones refuse food, preferring self-annihilation to disempowerment. And though they have been denied the acquisition of language – why would patented materials need to talk? – the surviving clones find a way to mourn the loss of their comrades by recreating an eerie Tibetan chant that they had heard on a factory-room recording. Finally, since active resistance is impossible, they form suicide clusters: groups of clones who calmly strangle each other until the last one left bangs his or her head on a cement floor – again, preferring death to exploitation. These are the resistance tactics that constitute powerful (though limited) forms of *réplique* – retorts or responses to the dehumanizing situation of the corporation's clones.

Such *répliques*, I will argue, are also at stake in our reading of *Orphan Black*. But before returning to that series, let me briefly discuss the word *réplique*'s multiple meanings – replica, replicant, reply, retort, revolt, aftershock – as they make their way through science-fiction writer Maurice Dantec's 1999 novel *Babylon Babies* (written in French, but after Dantec had moved to Canada). This cyberpunk novel is decidedly Deleuzian, and explicitly so: Deleuze appears in the book's acknowledgements alongside Donna Haraway, Philip K. Dick and punk rock bands like Aphex Twins and Fœtus (note the names' reproductive significance). Dantec dedicates his book to Eva, in an echo of Eve the first mother, and to 'les enfants du futur' ('the children of the future'), thus laying out a shift from biblical Genesis to '*neo*-genesis', Dantec's term for posthuman reproduction (1999: 11). Set in the then futuristic year of 2013, the novel tracks the nine-month gestation period of Marie Zorn, a young schizophrenic who merges with an AI network named Joe-Jane to become a '*maman-machine*' (Mama-machine), a Deleuzian 'body-without-organs' who is (paradoxically) reduced to *one* organ, the womb. This, however, is a womb redefined, by an age of bio-cybernetic experimentation with artificial placentas and altered genetic sequencing. Marie Zorn is the literal *matrix*, carrying within her biological womb two mysterious beings implanted there by a secret Kazakhstani lab while also serving as a carrier for Joe-Jane's prophetic musings on new forms of fecundity. Here is an exemplary, though shortened, extract from one of the *maman-machine*'s loghorrean flows:

> Rhythmic with blazing veins shooting into the future pure imminent cataclysm rising up like [...] the infinite primordial whirlpool inferno calling mankind beyond its permanent collision-computation-acceleration limits of impossible vaginal-spermatic-pollinating fluid of life in exponential profusion carrier wave of chaos [...] Marie Zorn is the zero-point for an ultraviolent explosion whose shock waves will be felt into the next millennia.
> (Dantec 1999: 393)

This prophecy of Marie as herald for a new reproductive age is developed over 600 pages, as the reader learns that what this new Eve has in her womb are neither (as hypothesized) transgenic mutant animals *à la The Island of Dr Moreau* (1896), nor a militarized virus sent to destroy North Americans through disease, but clones – more specifically, twin baby girls: two cloned humans who have been genetically programmed to fuse psychically with all human and supra-human knowledge of the universe. These babies, writes Dantec, are the 'posthuman' product of natural evolution and artificial technology: 'Homo sapiens neuromatrix, the Zorn twins are the beginning of the end of humanity' (Dantec 1999: 699).

One point worth noting in this techno-apocalyptic vision is that the foundationally fertile Marie Zorn is contrasted by Dantec with a 'Barren Bitch': the millenarian cult leader Ariane Clayton-Rochette, whose severe blond hairbun signals her asexual sterility. Ariane cannot bear children herself, so she plots to take Marie's clone twins to an interplanetary colony, where they will guarantee her immortality as leader of the Cosmic Church of the New Resurrection. (Again, I see echoes here with *Orphan Black*'s Rachel: sterility, coldness, will to power….). Dantec's frigid villainess fails, in part because Marie has been warned about Ariane's plot through a secularized 'Annunciation of the Virgin Mary'; in that scene, Marie finds herself in an inverted '*réplique*' of an apartment room in which the waves of a television screen send her an angel and its diabolical '*réplique*' to announce that multiple agents want to tear her twins, dead or alive, out of her womb (Dantec 1999: 388–89). Dantec's multiple use here of the term *réplique* brings me to a second point, in the form of a question: can the shift from biological reproduction to posthuman replication carry with it any sense of political leverage, of radical possibility for change in the world? In his depiction of Marie Zorn as *maman-machine*, Dantec cites Deleuze and Guattari's definition of the schizophrenic as the 'exterminating angel' of capitalism (Dantec 1999: 149). In this way, he proposes Marie Zorn as a *réplique*, a response or retort to the world as it's currently configured – but one that is much more violent than the clone rebellions we saw in Truong. For Marie Zorn's *réplique* takes the form of a brutal counter-attack or aftershock, an apocalyptic vision in which her twins (having resisted being torn out of the womb) will decimate humanity as we know it. Dantec's 'New Eve' is a singular matrix, in other words, of destruction as well as creation. In the end, his novel combines, somewhat paradoxically, a radical vision of non-biological reproduction with conservative religious politics. (Dantec is known for his adult conversion to Catholicism; one of the stranger passages in *Babylon Babies* has Marie Zorn come back after death as the Angel of Desolation floating through the Milky Way with a million star-souls of aborted children.) So, to recap: radical maternity, conservative politics. This combination in Dantec constitutes, I would argue, the opposite of what we find in *Orphan Black*. In the final section of this chapter, I will use the theories of Malabou and Goh to extend Truong's and Dantec's reflections on the reproductive and revolutionary capacities of clones in order to argue that *Orphan Black* mobilizes a rather traditional glorification of motherhood in the service of radically *new* forms of family and community that serve as *répliques*, strategic retorts to the dehumanizing efforts of the military–industrial complex.

Clone Club and the Biopolitics of Adaptation: Malabou and Goh

In her 2016 essay 'One life only: Biological resistance, political resistance', the philosopher Catherine Malabou emphasizes the *bio* in biopower to propose that living bodies might carry within them the potential to resist political regulatory systems striving to control them. Two technologies in particular reveal 'the reserve of possibilities inscribed in the living being itself': the first is epigenetic and '[t]he second is cloning, with its two fields of operation: asexual reproduction and regeneration (or self-healing)' (Malabou 2016: 431). Malabou's discussion of cloning allows us to distinguish usefully between the two main meanings of *réplique* raised by Truong and Dantec: replica and retort. She writes: 'The challenge that cloning makes to the category of difference is not related primarily to the copy, to the threat of an eternal return of the same, for the clone will never be a faithful and perfect copy' (437); in terms of *Orphan Black*, environmental variations allow the multiple clones to emerge as differentiated individuals. And if, as Malabou asserts, this differentiation raises the challenge of going back to 'a time *before difference*' – that is, a time of cellular regeneration rather than sexual reproduction – *Orphan Black* literalizes that return by identifying the 'chimaera' Kendall Malone as the double-sexed biological source of the Leda and Castor clones. 'The posthuman is thus also the prehuman' (Malabou 2016: 437). But if *Orphan Black*'s replicants are not mere replicas, they *do* carry within them the potential for *réplique* – in the sense of retort – that Malabou finds located, potentially, in the phylogenetic return to the prehuman state. 'Biological potentials reveal unprecedented modes of transformation', proposes Malabou; a 'conception of the self as a source of reproduction' and an opening up to the plurality of biological life can 'achieve a veritable deconstruction of program, family, and identity that threatens to fracture the presumed unity of the political subject' (2016: 438). Thus we arrive at a way to think of *Orphan Black*'s valorization of fertility, the clone's capacity for reproduction, as undercutting rather than reinforcing neo-liberal ideologies of coupled maternity. Remember, too, that the two bodies most intimately linked to Sarah Manning's maternal function – Kira's and Helena's – have an inexplicable capacity for self-healing: Kira defies medical understanding in her recovery from a car accident and Helena returns from apparent death by gunshot. They share, in other words, the capacity for cellular self-regeneration that Malabou sees as one of the 'possibilities inscribed in the living being itself' (2016: 431).

The questions that Malabou asks about these 'possibilities' are crucial: 'And how might the return of these possibilities offer a power of resistance? The resistance of biology to biopolitics?' (2016: 438). One tentative answer that Malabou proposes has to do with plurality. Thinking plurality would involve moving beyond bodily boundaries to find areas of trans-individual biological potential against the 'presumed unity of the political subject'. Malabou concludes her essay with a somewhat enigmatic but exciting vision of such transformative possibilities:

> The articulation of political discourse on bodies is always partial, for it cannot absorb everything that the structure of the living being is able to burst open by showing the possibilities of a reversal in the order of generations, a complexification in the notion of heritage, a calling into question of filiation, a new relation to death and the irreversibility of time, through which emerges a new experience of finitude.
>
> (2016: 438)

The temporal implications of Malabou's post-Darwinian vision are intriguing, as they seem to posit that cloning technology fuses the pre-sexual biological past with a future in which individual bodies may die but a certain version of the 'self' lives on. But more to my point, I think that Malabou's postulation of (clone) biology as carrying the potential for a radical rethinking of heritage and filiation has direct relevance for our reading of family, fertility and resistance in *Orphan Black*.

Perhaps even more helpful to thinking through the trans-individual potential for *réplique* in *Orphan Black* is Irving Goh's 2015 book *The Reject*, which proposes that term (the 'reject') as a response to Jean-Luc Nancy's question 'who comes after the Subject?' – i.e. *after* the totalizing self-presence that theorists from Cixous and Clément to Derrida and Deleuze have effectively critiqued. The subtitle of Goh's book is *Community, Politics, and Religion after the Subject*; through a rethinking of those categories, Goh aims to redraw the contours of ethical group relations in ways that pertain suggestively to *Orphan Black* – and in particular to the shifting, amoeba-like family cluster that is Clone Club.

In his book, Goh presents what he calls the three conceptual 'turns' of the reject. The first is 'the *reject* as it is conventionally understood: a *passive* figure targeted to be denied, denigrated, negated, disregarded, disposed of, abandoned, banished, or [...] exiled' (2015: 7). We might think here of Truong's clones in *Reproduction interdite*: hunted, cut up for body parts, disposed of as trash – or alternately, in *Orphan Black*, of the *pre*-self-aware clones, subjected to invasive experimentation on wombs and brains without consent or knowledge (think of BrightBorn), and eradicated when no longer useful ('Remember Helsinki'). These are passive victims. The second turn of Goh's theory reverses the power dynamic through the notion of retaliation, a forceful rejection of the external forces acting against the reject. That retaliation might take the form of nihilistic or violent acts such as, in *Orphan Black*, Helena's initial extermination mission in Season 1. Although she has misdirected her murderous rage, first against the Ukrainian nuns who abused her and then (manipulated by Tomas) against her own sister-clones, Helena is lashing out against the external forces that she understands to be acting against her; in this way, she inhabits Goh's second 'turn' of the reject. And her acts, like those of Goh's second turn, are also ineffectual: Helena's attacks go nowhere towards undoing the actual power systems of her world (Dyad Institute, the US Army, etc.) – in fact, she seems continually destined for a return to passivity (police captive, Prolethean brood mare, military lab rat), if not for the intervention of her *sestras*. And here, with the *sestras*, is where the third 'turn' comes in. Because the second turn (retaliation) risks falling back into the self-imposing logic of the subject, Goh proposes a *third* turn – called the 'auto-reject' – as the only truly effective strategic response to dehumanizing forces and systemic violence of State surveillance or corporate systems.

The auto-reject, writes Goh, turns the force of rejection around on itself as an act of creative regeneration, not self-annihilation. (In this way, it is unlike Truong's suicidal clones.) And it does so by taking a diagonal, secant trajectory, by making the choice of sidestepping or walking away from destructive systems in order to preserve life.

> Sometimes, this might involve a shift or a sidestepping to an adjacent space; at other times, it involves an adoption of an entirely other mode of *being*, which might also mean once again the renunciation of one's previous thought, actions and behaviour, and hence freeing the *reject* in another instance from any hypostasis of thought and action. The

> additional ethical force comes from the fact that all this stems from the affirmation and respect of others.
>
> (Goh 2015: 8)

From her very first encounter with the same/other uncanny of the (corporation-made) clone on the train platform in 1.1, Sarah Manning's survival strategies take these 'third turn' forms of sidestepping and of self-reinvention. An obvious example is her falsified suicide in order to take on the identity of policewoman Beth Childs in Season 1; and of course, she and her sisters will all take on each other's identities throughout the series, in continual re-adoptions of 'other mode[s] of *being*' (Goh 2015: 8). The specific strategy of sidestepping also recurs: 'walking away' is what Sarah does at the end of Season 2 for example, after the concerted pencil attack on Rachel's eye in the clinical operating room of the Dyad Institute. If Helena had been in the same situation – an escape from captivity and a threatened oophorectomy – she would likely have tried to lay waste to the place (as she actually *will* do at Henrik's farm); but Sarah, on the other hand, retreats and regroups. In fact, she literally regroups, reinforcing the solidarity of the proto-Clone Club, which in its pencil plan was not actually trying to destroy Dyad – it was just creating an exit path, a way out of the victim narrative for Sarah. This is what Goh calls the 'liberating trajectory' that allows the auto-reject to free itself from any fixed predicates without a forgetting of its traumatic past (2015: 10). As opposed to a Zizekian 'doing nothing' or abdication to *Violence* (Žižek 2008), Goh's auto-reject *acts*, and it does so through an ethics of solidarity that goes beyond existing *doxa* of friendship and community.

'Is it possible to think a friendship, love, or community of, and for, *rejects*? [...] And can the *reject* even be the figure of thought for a critical rethinking of contemporary friendship, love, and community?' (Goh 2015: 25). Goh answers his own questions by citing Jean-Luc Nancy's call to think of a 'community *without subject*' and, more affirmatively, Jacques Derrida's politics of friendship (*Politiques de l'amitié*) as non-hegemonic sociability. Derrida's thought provides for Goh a way to 'consider other rhythms and trajectories that seek to sidestep and resist the hegemony of hyper-gregarious network-centric friendship', thus allowing a community of rejects to avoid capitalist ideological appropriation and to dwell, instead, in the productive abyss or essential tear [as in 'rip'] in the thought of friendship (Goh 2015: 35–36). Such a friendship community would exist both outside of digital social networks (like Facebook – or the official, traceable Internet of *Orphan Black* that Season 4's hacker M.K. undercuts) and outside conventionalized articulations of affection. Here we might note that the deepest bond between Helena and Sarah exists not just despite – but also *because* – they have tried to kill each other. Think, too, of how affectionately Sarah calls Cosima a bitch, how Alison and Donnie are made stronger by their mutual distrust, or how Felix and Sarah continuously give each other the freedom to depart. 'As *rejects* in or before friendship, each and every one of us will have the right to refuse to respond, or even the right to walk away, silently, without explanation, and there will be neither rapprochement nor reproach' (Goh 2015: 46). (When Cosima signs a contract with Dyad, Sarah simply replies: 'no judgment'.) If Clone Club feels both new and authentic to the viewer of *Orphan Black*, it may be because its members inhabit the paradoxical double stance of radical autonomy and radical solidarity. In his exploration of new constellations of auto-reject community, Goh also cites Catherine Clément's 'syncopic love', Luce Irigaray's 'walking away' and Deleuze and Guattari's 'always already secant' notion of communal relations as having mistrust at their core; the force of rejection must be at the heart of friendship for there to be the radical freedom of the other. These philosophical notions of community after the *subject* are perhaps difficult to imagine being put into practice. But I would suggest that through the shifting, transient, constantly adaptive contours of Clone Club, *Orphan Black* allows us to envision the implications of a radically non-saccharine response/*réplique* to conventions of friendship, of community bond, of family.

After all, Clone Club is all about redefining family. The connection need not be biological (Helena on Felix: 'he is *sestra*?'), but it does entail an ethical imperative: in 3.3, Sarah says to Mark, 'you have to care about Helena, she's your sister' – and later, to Gracie, 'you have to care about Helena, she's the mother of your unborn child'. The redefinitions of family are made especially salient at the end of Season 3: when Mrs S says to a pregnant Helena, 'join us, you're kin'; when Helena tells drug dealers (before offing them) 'you should not threaten babies'; or when Cosima helps Alison make an election speech in which she says 'if the family suddenly gets bigger, you make room. You adapt'. Adaptation, this Darwinian logic that both subtends and is bypassed by *Orphan Black*'s 'posthuman' clones, is the key to survival. Survival – the continuation not of the older terms of individual and species but in newer terms of community (Clone Club) – implies reproduction, regeneration. And regeneration, in the case of *Orphan Black*, implies motherhood. Sarah's survival strategies are fundamentally motivated by protecting Kira. In opposition to 'bad mother' Dr Coady, destroyer of procreation, Mrs S represents the maternal transmission of tactics for survival; 'if raising Sarah has taught me anything', she says, 'it's how to survive worry: you keep on living' (3.6). But Mrs S is no namby-pamby apron-wearing typical mom. Her IRA training and occasionally hard heart remind us that reproductive rights in *Orphan Black* are more than just a centrist 'Planned Parenthood' type of political stance; they are based in a violent world of (clone) alterity and are thus aligned, rather, with Goh's theory of the *reject*, in which the goal of secant or transversal strategy as *réplique* to the military-corporate system is survival. As we saw earlier, in Truong's novel *Replication interdite* the sterilized clones' response to oppression is severely limited; they can only replicate sound (Tibetan chants), but they cannot reproduce – or survive. And in Dantec's *Babylon Babies*, the *réplique* born of Marie Zorn's womb is singular and apocalyptic, with no clear vision for a future community. But *Orphan Black* gives us, with its posthuman community of 'reject' clones, a possible model for the survival of our current neoliberal condition.

The Clone Club family has allowed Helena to move from murderous angel of fertility in Season 1 (she killed Katja because she was 'unfit for family' and threatened Sarah with a knife emblazoned with the Prolethean fish symbol of fertility) to protective mother of embryos developed in a canister in Season 4, thus wresting the very notion of reproduction from the

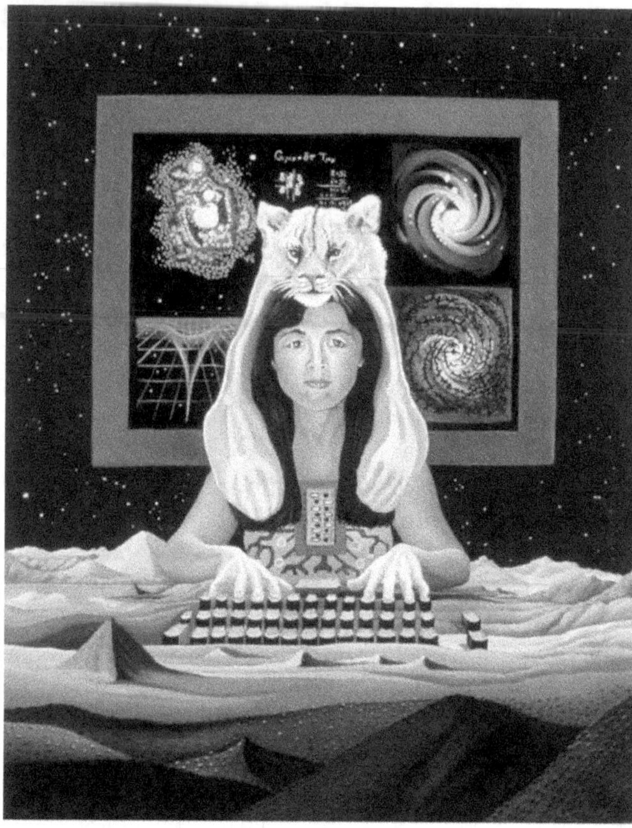

Figure 26: Lynn Randolph's *Cyborg*.

realm of perverted religion to that of science, perhaps even to that of the new materialism to which Malabou's neuroscience-based philosophy point us. Fittingly, the series' third season ends with the open possibility of new births and new notions of motherhood – whether biological, as for Sarah and Helena; or not, as for Rachel, who during the season finale is given her younger clone-sister Charlotte by her own adoptive mother Susan Duncan, now head of Neolution, the ultimate matrix of posthuman reproduction. And in Season 4, whose episode titles move beyond the Darwinian quotes of Season 1 to citations from feminist critic Donna Haraway's 1991 *Simians, Cyborgs, and Women*, we shift again to a trans-individual notion of biology (and biological motherhood), as the *sestras* move in and out of each other's lives and parenting roles with amoeba-like fluidity. Even the cover image of Haraway's book, a 1989 painting by Lynn Randolph entitled *Cyborg*, emerges as a potential visual intertext for *Orphan Black* (Figure 26): a hybrid woman engages with the world around her through a computer keyboard (like M.K.), with a bearskin covering her head (like Helena, in her primitivist back-to-nature stance of Seasons 4 and 5), an attractive feminine hairstyle and face (like the more conventional, yet increasingly strong, Alison), a scientific chart of the universe (like Cosima, whose mastery is based on reason and research), an implanted technological chip in her chest (like Rachel, now a true 'cyborg' with her artificial eye) and a panel in the abdominal – or womb – region that seems to combine network technology with the throbbing life-blood of neurology (like Sarah, the *mother* at the centre of Clone Club). By combining feral primitivism with cybertech prowess into one trans-individual image of female empowerment, *Orphan Black* mobilizes the possibilities underlying Haraway's feminist 'cyborg'.

In 3.1 as Sarah is struggling to evade Topside's evil threats and to figure out with Cosima whether Helena is being used as a pawn, all the while haunted by the terrifying spectre of the eradication of clones in Helsinki, she decides: 'We have to make our own moves'. These are the 'sidestepping' moves made possible by the new form of family that is Clone Club – an 'affinity group' (Haraway 1991: 155) or hybrid trans-individual entity that can mobilize reproduction into resistance, regeneration into *réplique*.

Afterword: Reflections on the Show, and Interviews with Cast, Crew and Creators

Lili Loofbourow

If it's broadly true that *Orphan Black* is *Jurassic Park* for 'clones' (see Porzak's essay in this volume) – a parable about the risky intersection of science and business, featuring a clutch of genetically modified females whom their hubristic creators expect to control – then the television series is much less about 'nature finding a way' than the messy project of marrying nature to technology and studying the resultant 'family'. That science-fiction is a lens through which to explore increasingly urgent questions – here, of female bodily autonomy – is a rhetorical cliché; with *Orphan Black*, however, one of the show's trademark pleasures is that the 'lens' in question becomes shockingly literal and embodied.

Take the moment when Rachel Duncan pries Dyad's prosthetic eye out of her own skull in Season 5's 'Gag or Throttle' (5.7). It's an act almost as redemptive as it is horrifying, and so much symbolic weight has accrued to it by this point that it practically throbs with narrative resonance. Viewers will recall that *Sarah* first stabbed Rachel's eye with a homely No. 2 pencil – just in time to escape a forced oophorectomy in 'By Means Which Have Never Yet Been Tried' (2.10; see Figure 28); this time, it is Rachel who voluntarily re-mutilates herself, shortly before she's supposed to harvest and deliver Sarah's daughter *Kira's* eggs. The formal symmetry to this pattern – removing eyes instead of eggs (see Goulet's essay in this volume) – gives it the confusing properties of a two-way mirror. In that bloody sacrifice of her vanity, her vision and her 'father', Rachel demonstrates that cures are just as bound up in questions of bodily autonomy as other (perhaps less welcome) interventions.

The show doesn't always plot out this problem as clearly as it could: Cosima's desperation for a 'cure', for instance, makes fine emotional sense, but given what she knows about Dyad/Topside/Revival by later seasons, her willingness to accept whatever treatment the Revival folks (and Rachel, in particular) deem appropriate is stunning – and blurs the boundary between trust and naiveté. But even in these hard-to-swallow scenarios, consent remains the crux: Rachel *offers* to inject Cosima with her 'cure', and Cosima *accepts*. Rachel wasn't afforded that courtesy, of course. Her eye was replaced via Neolutionist wizardry through the kind of emergency measure we tend not to think of as requiring consent – at the time, it didn't seem to matter that the eye was implanted without Rachel's full participation (she was unconscious, after all, and had suffered brain damage).

But consent always matters in *Orphan Black*, and the significance of that prosthetic implant has multiplied over time. Narratively, it hearkens back to the Neolutionist affectation of sporting an all-white eye in Season 1, when the Neolution nightclub hostess hopes that this cosmetic symbol of technologically enhanced perception would one day be fully functional. (Aldous Leekie displays just such an eye in a presentation in 'Variations under Domestication'.) In Rachel, that 'vision' has become a reality.

That the eye does much more than Rachel supposes is precisely where the show militates against simplistic analogies between natural processes and their technological counterparts, replacements or supplements. As it turns out, the prosthetic is not just a cure, not just an eye: it's a camera. While it approximates a human eye's limited functions, it does a great deal more besides. It becomes clear in the fourth season that it also acts as a delivery system for P.T. Westmorland's projected images, including the swan representing Susan Duncan's doomed side of the Leda project. Rachel, it turns out, was effectively outfitted with the world's most invasive film projector. That Rachel was initially both intrigued and converted by these films isn't particularly surprising; as the show's happiest watcher of home movies, whatever limited sense of familial connection she has stems from exactly this kind of spectatorship. But she discovers later, in Season 5, that the prosthetic had also, perversely, transformed her into *her own monitor*: it was transmitting footage of everything she saw back to P.T. Westmorland. The mirror reflects two ways.

This transformation of the self into a panopticon is a shattering betrayal from a self-proclaimed father figure who seemed determined to keep a promise broken by Rachel's other 'fathers', Ethan Duncan and Aldous Leekie. Westmorland had just proclaimed Rachel legally and functionally 'emancipated' – a full and equal participant – when she discovered the secret screen where he would forever watch her life's footage and weaponize her attachments. His apparent fatherly insights weren't just feigned, they were filched.

Because the clones are largely fatherless, it makes sense that Rachel's desperate drive to differentiate herself from the rest would result in her being awarded three by those charged with managing her.[1] But by exposing Westmorland and destroying her eye, Rachel finally rejects the story of her exceptionalism. For most of her life, she thought she alone saw clearly; as the sole self-aware clone, she believed herself capable of a full and privileged objectivity that her fellow clones lacked. But the prosthetic's function as a camera condemned her to an untenable subjectivity – and to permanent status as a *subject*. The poetry of its destruction, then, is rich. She revealed her 'father's' identity once it became clear that she remained more valuable to him as data than she ever would be as a disciple, let alone a daughter.

(There are also, of course, Biblical echoes to plucking out your eye if it causes you to stumble: Matthew 5:29. That the scriptural injunction refers to infidelity is more than fitting; we can debate the scope of Rachel's 'daddy issues', but what's certainly clear is that she's being unfaithful to her 'father' by plucking out her eye – and desiring something he doesn't want her to desire.)

That the story of Rachel's prosthetic lens resonates on these many frequencies captures *Orphan Black* at its best. At its peak, the show hybridizes bioethics with drama in ways that deepen both – generating a theoretical superstructure that lends the series a formal elegance anchored by real human stakes. But a show with this much plot can also get clunky and overburdened; the story of Rachel's eye is as structurally tight, symbolically resonant and dramatically rich as the maggot bot stuff in the fourth season was messy and inert. Evie Cho's horrifying grubs shared the prosthetic eye's potential for mission creep (they too constituted a kind of Trojan horse, which, once installed, could achieve any kind of 'gene therapy' the Neolutionists wanted); if people's genomes became editable and subject to Neolution's control, Evie Cho wouldn't *have* to limit her experiments to BrightBorn's genetically engineered babies. But while Cho's bots were a perfectly adequate villainous conceit, the comparative shallowness with which that season played out is instructive: the bot metaphors didn't mix or play off the symbolic potential of Kendall as a 'chimera' or 'original' with anything like the productive echoes between Rachel's eye and Sarah's eggs. That Cho killed Kendall felt less than organic as a result. Given that it is later revealed that Westmorland's Revival experiments were happening during this whole subplot, this seemed more like a manufactured clash of corporate interests than a robust narrative arc.

Nurtured vs. Unnurtured

One possible reason for that failure is that Cho and Coady are villains disconnected from *Orphan Black*'s investment in social as well as genetic definitions of family. They clashed with Kendall's inclusion, which twisted the adoptive clone family into a biological pretzel: that Sarah is a clone of Siobhan's mother puts them into an unsuspected genetic relationship with each other. This is an interesting riddle, of course. But does it *mean* anything? Change anything?

It's unclear. *Orphan Black*'s concept of how genetics intersect with love is more volatile than it could be. For a show this conversant in the ways genetics get overread, there are plenty of moments when DNA appears to signify more powerfully than it should (see Neroni's essay in this volume): Helena, for instance, is emphatic about the deep maternal feeling she feels towards '[her] babies' on the basis of nothing but shared DNA. The 'magical connection' Helena feels towards Sarah is allegedly a function of their time in the womb together (though the fact that this isn't reciprocated by Sarah – who discovers a similar connection to *Kira* as she's trying to escape the island – is puzzling). As for Kira, her psychic connection to the clones never gets scientifically explained or emotionally realized, nor does that open channel seem to affect her particularly – she mentions an 'empty space' after Ferdinand kills M.K., but it's never mentioned again. These are all factors that implicitly credit biology with engendering an almost magical connection.

On the other hand, the show also boasts a gamut of alternative family structures. The array of nonbiological mothers includes Kendra, the beleaguered BrightBorn surrogate trying to save an infant born blind; Amelia, who carried Sarah and Helena, and seemed to harbour sincere affection for them; Susan Duncan, whose cool rejection of Rachel is hard to read in maternal terms and Virginia Coady, the Castor 'mother' whose experiments hardly qualify as parental. There's also Siobhan, of course, a tough but immensely affectionate mother, and Alison, though her children only sporadically seem to exist.

Siobhan's death forced the show to sharpen these definitions. When we were first introduced to Sarah Manning, Siobhan's brood hardly constituted a unit; Sarah had been away from her and Kira for months, with only Felix acting as a hapless ambassador. But if one of the show's pleasures has been watching these family structures solidify, their governing assumptions remain murky. Cosima's solemn assertion (in 'Violation under Nature') that 'we are your biological imperative now' seems to argue that DNA forms an instant basis for community. It's a strong statement from a show that includes biological as well as adoptive siblings and children, and which rightly regards genetics as less than determinative. That the show briefly featured Tony, a trans gender clone, demonstrates how badly it wanted to make this point, as does the extent to which Felix is treated as a '*sestra*', or member of the Clone Club. But it remains troubling that Kira is almost magic, thanks to her genetic links, while Alison and Donnie's adoptive children, again, barely rate a mention.

The first interesting test of how these guidelines work when strained comes when Sarah takes Felix and Kira to Cal's cabin. Felix bridles at being ambushed in this way and announces he's leaving; 'there's no place for me here', he says (2.3). The implications are interesting: as a queer adoptive character, he can tolerate and belong to the genetic sameness of Clone Club only until the traditional nuclear family conceptually locks him out. The second test arrives with the Castor clones, whom Sarah repeatedly tries to address and treat as 'brothers'. That attempt at connection never builds up to much – even Ira remains pretty distant and remote. Their shared genetic emergency fails to produce anything like real camaraderie. The third test comes when Adele, Felix's genetic half-sister, realizes that her closeness with him is – if not feigned – superficial: 'Genetics doesn't really make a family, does it?' she says when Felix refuses to tell her the truth about his 'sisters', and adds that what's required is honesty and acceptance. 'I'm going to leave you with your real family', she says (4.9). It takes Siobhan – Felix's adoptive mother – to hunt Adele down, fill her in, and recruit and induct her (via 'honesty and acceptance') into real kinship. It's telling, incidentally, that Sarah responds as poorly as she does to Adele's appearance in their lives. Her concerns that Adele might be a spy are understandable; the show rarely explores the paranoia someone living like Sarah would find routine, though this moment comes close. But there's a jealous undercurrent on Sarah's side, a concern that Felix might have a genetic connection with someone that trumps the non-genetic connection he and she share.

In fact, the clearest statement of what family means on *Orphan Black* comes from Felix, the genetic odd man out whose speech at his art show in 'One Fettered Slave' foreshadows Siobhan's death so loudly it's hard to hear much else. 'We could have ended up anywhere', he says of himself and Sarah; 'we could have ended up in any family, and if we had, we would have ended up as entirely different people. But my mum, Siobhan, this woman: she chose us as her own. We are who we are because she carried two little London urchins on her wings to Canada'. It's a lovely thing – the Leda reference, the emphasis on environment over genetics. The show is less than subtle about these things, and Felix goes on to make precisely that point: 'We are mysterious works of chance', he says, 'a choice of nature vs. nurture. So, to my galaxy of women, thank you for the nurture'.

It's a poignant moment, and a more thorough survey than this one would carefully analyse the moment when Sarah and Felix started calling Siobhan 'mum'. But in a show obsessed with eggs, origins and ontology, it seems right that Siobhan's last word before she dies is *chickens*.

As a counterpoint to these weak forces that result in something cohesive enough to be called *family*, there's value in considering how the orphans who grew up *without* such a group developed secret affinities of their own. Helena and Sarah tend to be grouped together because they came from Amelia's womb, but in many ways Helena and *Rachel* are the more interesting pair (see Figures 27 and 28). Aesthetically, of course, they're polar opposites: one is groomed, the other wild. One is refined, the other coarse. One is subtle, the other blunt. But each was raised by an institution. Each is sporadically sociopathic.[2] Each is responsible for the deaths of multiple clones. Each is capable of astonishing self-mutilation. Finally, and most interestingly, each gets shuffled through the show's good and evil sorting system and crops up on different sides of it depending on the show's generic needs.[3]

Rachel reacts to being left out the way Helena reacts to threats to children. And if it's Rachel's fate to *become* a camera – with films playing in her head – it's Helena's fate to become a film. *Orphan Black* is a mishmash of genres, and while Rachel's toggles between corporate crime and medicalized rape, Helena's is almost always a quasi-gothic horror flick. What distinguishes them is their ability to flip back and forth between being monster and the victim.

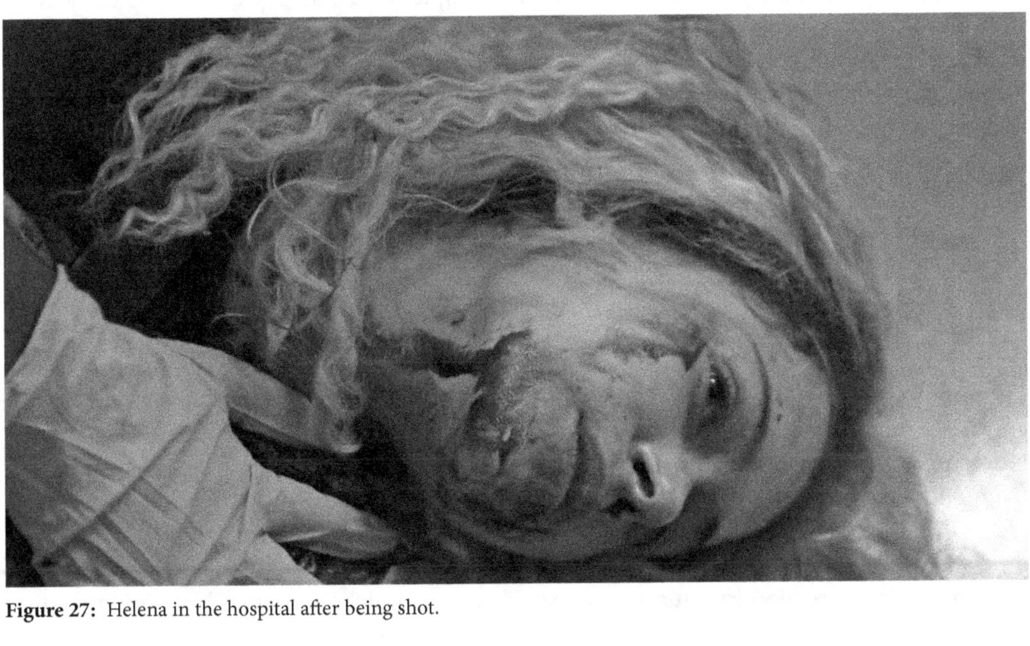

Figure 27: Helena in the hospital after being shot.

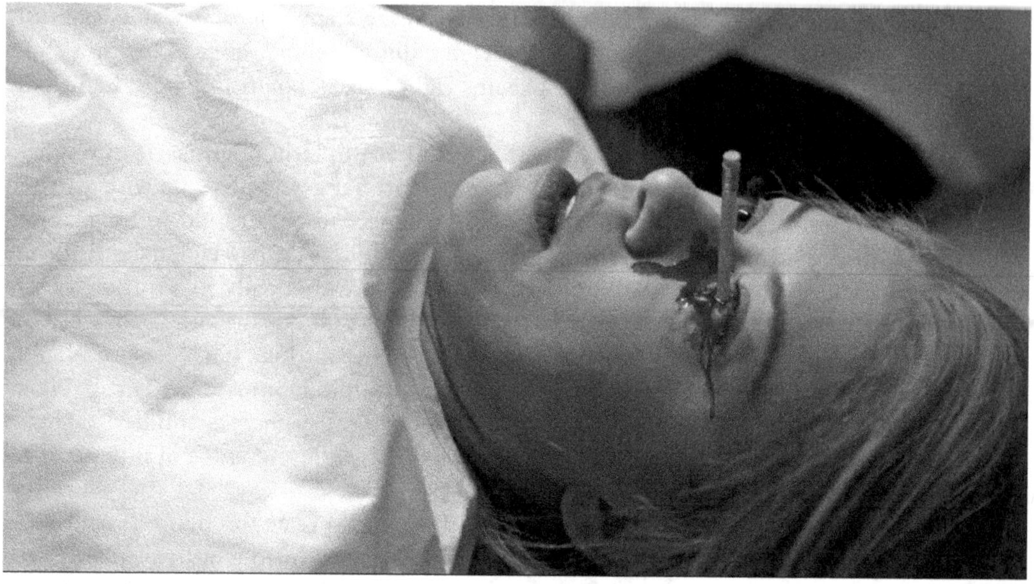

Figure 28: Rachel in the operating room after taking a pencil in the eye.

Sometimes, of course, they're both. But in no case do they permanently or fully conform to the expectations their types establish. Take Helena here in Figure 29:

This remarkable shot offers a glimpse of how simple it was, in *Orphan Black*'s early days, to overread Helena's scars and history of trauma. Her early appearances are terrifying and drenched in familiar symbolism: it's arguable, for instance, that she 'unrapes' herself by pulling the phallic iron bar Sarah stabbed into her abdomen, and that she sews herself up as a parody of conventional femininity. These are perfectly reasonable readings if you know nothing else about Helena and try to read her through the genres her scenes reference. But later in the show – once we know about her shadow puppets, her gleeful revenge-rape of Henrik and her clear-eyed acceptance of her pregnancy – those readings lose some persuasive force. Helena remains frightening, but her humour and cussedness prevent her from being the sublime archetype of suffering she once was.

There are a plenty of other horror devices at work in Helena scenes too: there's the cutting, the obsession with being 'The Original', the shrieky soundtrack, the jump scares, the imperviousness to blood. Not to mention the creepy doodles, the beheaded Barbies, and the sublime attitude she assumes when she's in prayer. I've called Helena TV's most redeemable female villain, and I stand by that: she's much too funny and lacks several of the qualities that typically define monstrous women. She's desexualized, for one thing, even when she's pregnant. Helena spends a lot of time covered in blood, her own and others, but for a show built around the intersection of technology, religion and female reproduction, it's remarkable that the blood in question is never, for example, *menstrual*. No *vagina dentata*

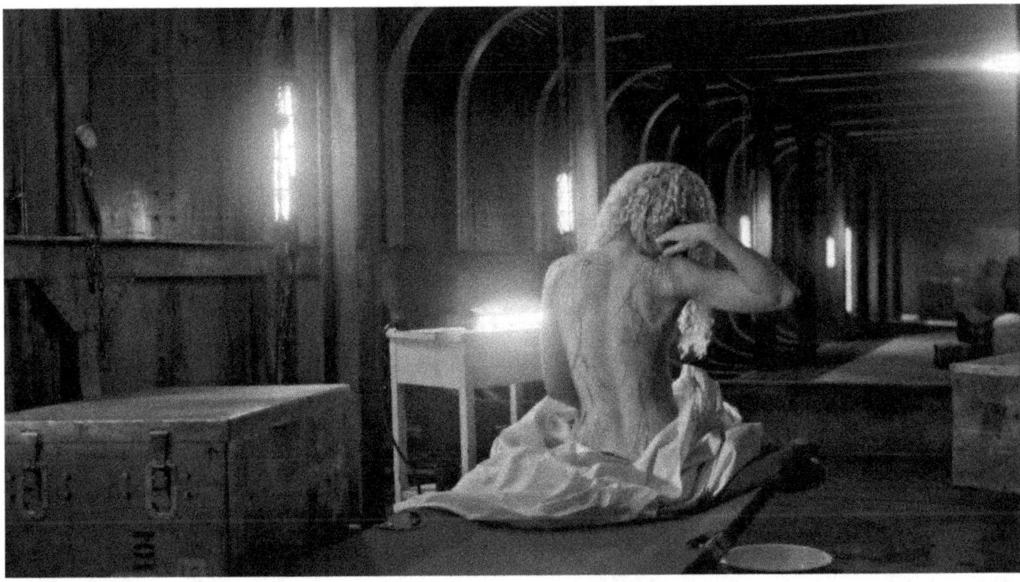

Figure 29: Helena's elaborate scars as a somewhat stereotyped symbol of trauma.

here, no fetishized virginity, no *American Horror Story*-like witches killing boys through sex. Helena is a monster whose superpower isn't her femininity but her ability to occupy both sides of the horror script.

That Helena initially fancies herself 'The Original' might be a sly wink to (among other things) Walter Benjamin and the scepticism that plagues a reproductive code when it starts to trump hard copies. There are other fantasies she explodes, the main one being the fact that monster and victim are stable categories. In *Orphan Black*, the victim becomes the monster becomes the victim again with bewildering and humorous ease. Just when you think you've settled, the camera cocks its head, and boom: something horrible happens. (This is as true of Rachel as it is of Helena.)

When *Orphan Black* has Helena in monster mode, for instance, it tends to shoot her boots-first. We see her army boots, and then we see her doing something awful: stitching herself up, or stalking someone, or about to attack. But the show's most interesting boots-first scene starts in a hospital corridor in 2.1, 'Nature Under Constraint and Vexed'. We thought Helena was dead from Sarah's gunshot. She isn't. There are the boots, dragging on the floor. There's blood on them. We hear the Helena soundtrack, and by the time the camera pans up to the monster-Helena shot, she looks like an actual zombie. This is what monsters do: they're defined by their boots and their blood, not their feelings or eyes. They're horrifying spectacles for us to look *at*.

But then Helena falls to the ground and the camera roguishly and redemptively slides up from her boots, those icons of horror, *to her face* (see Figure 27). And then – here's the connection to Rachel – the camera goes inside her eyes. Rather than 'other' the monster by staring *at* her, we see what Helena sees: first the blur of a helpful nurse, then Mark walking down the hallway. Her gaze focuses, and then, *horror*: there's a close-up Mark's belt buckle. Through a simple change in perspective, Helena has become the victim in the horror flick instead of the monster.

This would be satisfying enough if it marked a permanent flip from 'bad' to 'good' or from 'monster' to 'victim'. But that's not what *Orphan Black* does. The camera *keeps* flipping. The lens won't let Helena rest in non-monstrosity any more than it lets Sarah Manning stay in one place. When Gracie tries to murder Helena, for instance, the scene opens using all the conventions of horror in ways that frame Helena as the hapless victim. She slumbers peacefully in the foreground while Gracie the monster approaches. During the struggle that ensues, the camera remains on Helena's side. But then Helena stops struggling. Gracie thinks she's won. Enter the horror trope whereby you poke the villain just to make sure she's dead. To do so, we enter Gracie's point of view. The camera shifts allegiances, Gracie walks away via a camera angle we are all trained to recognize as a very bad sign indeed, and then: JUMP SCARE. Monster Helena is BACK.

She doesn't last long, though: soon after, Helena is running through hallways, dressed like a gothic bride and hiding like prey. And yet the first shot of her escape reminds us that she's got something important going on under her virginal wedding dress: army boots. By the end of this story, she'll have burned everything down.

Insides and Outsides

I began this essay by remarking on the fact that both of Rachel's enucleations were preceded by planned nonconsensual egg extractions: first Sarah's, then Kira's. Helena's flashbacks as she flees the Proletheans reveal that this exact procedure was performed on her: she was subjected to a nonconsensual egg retrieval.

Orphan Black is filled with disturbing funhouse mirrors like these. Cosima gets to study her own genetic material and discovers a patent is written into her very biological code; she doesn't own herself. Helena and Sarah Manning are mirror images of each other, right down to their organ placement, and these facts are plumbed for epistemological significance. Helena was taught by 'Tomas' to interpret her physical and genetic sameness with the clones competitively: to see the clones as evil copies, as 'sheep' that had to be slaughtered.[4] The Barbie dolls were an early gesture to this idea of women as generic, reproducible, competitive and interchangeable, metacommentary on the bland sameness on offer when it came to roles for women. In my 2015 profile of Tatiana Maslany I argued that, by incorporating so many different styles and types, *Orphan Black* talks back to television's tendency to maroon lone women in their respective genres by placing those genres (and the women in them) in relation to each other. I'll briefly revisit what I said then, because a small but significant correction is in order:

> In its subject matter, 'Orphan Black' broods on the nature-nurture debate in human biology, but in its execution, the show cleverly extends the same question to matters of genre. What does the exact same woman look like if you grow her in the petri dish of 'Desperate Housewives' or on a horror-film set in Eastern Europe? What about a police procedural? The result is a revelation: Instead of each archetype existing as the lone female character in her respective universe, these normally isolated tropes find one another, band together and seek to liberate themselves from the evil system that created them.
> (Loofbourow 2015)

When I interviewed Tatiana Maslany for that piece, I mentioned that the main figure missing from *Orphan Black's* galaxy of women was the typical rom-com heroine. 'Oh yeah', Maslany said then. 'Yeah, I think Jenny Slate talked about her. The girl who like can't keep her purse on. "Mmm, it fell off for no reason"', she says, doing a decent impression of Slate's PubLizity character in *Kroll Show*. 'The quirky girl' (Maslany 2014).

That's exactly it, and I might have added that there was no 'Barbie' clone – no purely vain blonde for the show to mock and ogle. The punitive standards of the industry to which this show belonged had somehow escaped comment. But if those once seemed like interesting omissions, they no longer obtain: Krystal Goderitch – a beautician and YouTuber clone who shares some features with Elle Woods of *Legally Blonde* (2001) – supplies those missing links in *Orphan Black*'s panoply of femininities. To the show's credit, it elevates its most apparently vapid character into a semi-successful sleuth, and Krystal's amusing investigations yield more than anyone suspected. She's *right* about 'Big Cosmetic', there *is* a 'dermal delivery system', and BluZone Cosmetics *was* just purchased by Dyad. There's a certain brilliance to Dyad expanding so as to control women's outsides as well as their insides.

Behind the Camera

There's a tendency, in the readings I've offered here, to treat the show as a unitary text. But no responsible afterword about a television show could ignore the complicated reality of television production: namely, that it is collaborative, messy and subject to so many external factors that it won't always conform to any unified artistic vision. Take the Krystal clone: in addition to all I've said about what she means for the show's careful survey about women's representation, her whole cosmetic subplot also doubles as a simpler inside joke about the show's dependence on *really great* make-up that assists Maslany in making the clones clearly and recognizably distinct; another meta-joke Krystal makes possible is that her target, BluZone Cosmetics' CEO Len Sipp, is played by Maslany's real-life partner, Tom Cullen. These sorts of games abound in television production, and having considered the lens and the camera from an analytical and symbolic viewpoint, it's useful to take another step back and think very literally about the nuts and bolts of television production. How does the story actually get written? How does the stuff happening behind the *real* lenses and *actual* cameras complicate the foregoing efforts at interpretation? How do television's more collaborative and spontaneous dimensions affect a show this precisely calibrated and technically difficult?

Let's start at the beginning. The show's co-creators, Graeme Manson and John Fawcett, met at the Canadian Film Centre, and have been working together for almost two decades. Here's Manson describing the genesis of *Orphan Black*:

> When we were thinking about clones and the kaleidoscopic maze that could be really fun for storytelling and images, John thinking as a director about the challenges and the fun of doubling and the switcheroos and the surprises, and myself immediately questions of theme and nature and nurture and the interesting side of clones that hasn't really been explored on TV before – the human side.
>
> (Manson 2014)

They split responsibilities along similar lines when the show began production: 'We basically work together on the big picture story elements – creating the character arcs and the season stories', Fawcett told me, 'and then [Graeme] basically runs the writers' room. He and the rest of the writers work out the details'.

> I'll be involved in that process as much as necessary. Sometimes it's better though for me to be a little bit peripheral so that I can be a little bit objective, as they bring the story pitches in, and offer feedback to the writers and to Graeme in that capacity. From my end of things, I will run the post-production department and supervise the editors and composer and visual effects and sound guy, and keep Graeme a little bit – not have him completely absorbed in it so that he can keep a fresh eye and be objective when looking at edits and cuts of the episodes.
>
> (Fawcett 2014)

So, what about the clones? Graeme and Fawcett based specific clones on people they knew. It's well known that Cosima is based on Cosima Herter, Manson's friend and the show's science consultant. Alison shares some features with John Fawcett's sister (Saraiya 2014). But Helena's creation was more complex, and the story of how she came to be explains some of her inconsistencies and some of her magic.

Helena's aesthetic was largely inspired by religious iconography. Stephen Lynch, *Orphan Black*'s versatile make-up artist, borrowed from *The Passion of Joan of Arc* (1928) for her look – 'such beautiful imagery, and the *pain* in her face', Lynch told me (when I spoke to him, he was preparing to paint the colour into Helena's prosthetic scars). There were other influences too; 'the Somerset Maugham movie with Bette Davis and Leslie Howard', he said. 'A syphilitic Mildred, eyes ringed'. He mentioned Giotto, and apologized for the pretentiousness (Lynch 2014).

Co-creators Manson and Fawcett had envisioned Helena a little differently; they'd planned for her to have angel-wing tattoos. Lynch opposed this on principle; to make his point, he canvassed the background holding area where the extras were waiting. There he found not one but *two* college frat boys with angel-wing tattoos. He showed them to Manson and Fawcett, telling them 'we've gotta do something better'. So, he went home and picked up a Sharpie and started 'cutting' himself with it. 'I did it on myself and I thought, if this were a knife or a razor—and so I drew on my own back, and I had a friend take a picture of my back, and I copied it a little bit, and I added to make it look more like feathers. That way it's scarification, so she can recut' (Lynch 2014).

Lynch thought a lot about the *lack* of make-up on *The Passion of Joan of Arc* as he came up with Helena's look. 'You wouldn't know it's a silent film', he tells me. 'It's the lack of make-up, I think…' (Lynch 2014).

John Fawcett independently offered a very similar account of Helena's creation:

> Stephen, in the makeup department, had this idea of cutting. We liked it, because it sort of suited her weirdness and her strange religious upbringing and the abuse that she'd suffered in childhood. It worked really well from a character point of view. I think when people – not just actors, but crew – when people are invested in a show, and they feel their ideas matter, I think that's when you get an environment of creative excitement, you know? I think both Graeme and I are collaborative dudes. That's just kind of how we roll. We know what we want, but you never know where you're gonna get the juice from. (Fawcett 2014)

As for Helena's peculiar and intense love for Sarah? That came from Maslany herself. Alex Levine, a writer on the show, explained that 'she was the one who, when we first started trying to figure out who this character was, and how she felt about Sarah, she said "I think she comes from a real place of love for this other person"'. It wasn't how they'd written the character, but it gave them pause: 'we realized it was a much more interesting choice', Levine said (Levine 2014).

Levine also wrote the episode where Helena revenge-inseminates the Prolethean leader, but he claims no credit for it: 'They ad libbed that, by the way', he tells me. 'The general idea of the revenge was there, but a lot of the dialogue in that scene – she and Peter ad libbed – yeah, we were running out of time there, shooting handheld, and T.J. was like *go*, and *yes, absolutely*' (Levine 2014). Maslany confirms: 'What did I just do *to* a man in front of 40 men?' she says she was thinking (2014).

I offer this rather complex Helena genesis story as an instance of the debates, processes and compromises that go into the creation of a character. Consistency won't always be the result. Helena is a little too funny to be either a monster or an angel, and the show riffs on the dissonance between Helena's occasional sublimity and her habitual raunch. She's also the product of multiple imaginations, and some of her inconsistencies are just that.

Maslany said her process for developing the clones initially required a lot of writing:

> It was a lot of, like, kind of brain work. And research and books, I had loads of books I read for all the characters and films that I watched. But then after that it became just

> keeping it in my body and making it less so about an intellectual thing because I just kind of had to trust it was all there.
>
> (Maslany 2014)

She also worked from the outside in; Maslany credited the hair and make-up team with developing looks that in many cases defined the character: 'Complete artists', she said.

> I couldn't have done it without them, and I couldn't do it in the future without them. Cosima's tattoos or nose ring, Alison's bangs, just these things that were instant keys that changed how I looked at myself. We can't help but be affected by how people see us and how we see ourselves, physically, aesthetically, you know what I mean, we're so, um, aesthetically driven in our culture, um, and uh so much too about the psychology of the aesthetic informing or negating what's actually going on inside. That was really exciting to me, because Alison's hair? There's not a thing out of place. But really inside, she's *I'm totally screwed, I don't know what I'm doing, I don't think I'm a real person*. I like that conflict.
>
> (Maslany 2014, original emphasis)

Here's Maslany's rundown of how she'd come to think of the different clones' personalities: Cosima 'lives in a world where everything's fascinating to her. The world is exciting, it's tactile. She wants to understand everything, she wants to be inside of people'. She saw Sarah as Cosima's opposite: 'She grew up never knowing her family, and has consequently very little trust for the world around her. I think so much of that indicates how you walk in that world, so Sarah's got a tension in her, a kind of boyishness almost'. As for Alison:

> a bit of a slow burn. Some people hated her. She's a type. I think, you know, I think she is terrified that her world's going to fall apart, and so her rigidity is in her body. I think she was a ballet dancer. I think all of her habits come from ballet, and I think she still holds her butt this tight and her stomach in, you know, just because she's proper. And she doesn't have any ill thoughts and she doesn't have any sexual feelings and she doesn't want you to know any of that.
>
> (Maslany 2014)

Maslany tells me all this at a table we share with the show's publicist, and the experience is a little vertiginous: we're effectively being monitored by a giant corporate interest. The irony doesn't end there: Maslany was careful to connect the show's concerns to her own experience of fame. '*Oh*, this is about volition and autonomy!' Maslany said, channelling her reaction to watching the first few episodes for the first time, 'that was resonating with me as far as being an actor who was being interviewed and being dressed' (Maslany 2014). She'd repeatedly described the peculiar experience of fame as something that can't be allowed to define you, but by interviewing her, I was contributing. I was arguably a monitor myself: someone staring at her, documenting her and reporting back to a curious public.

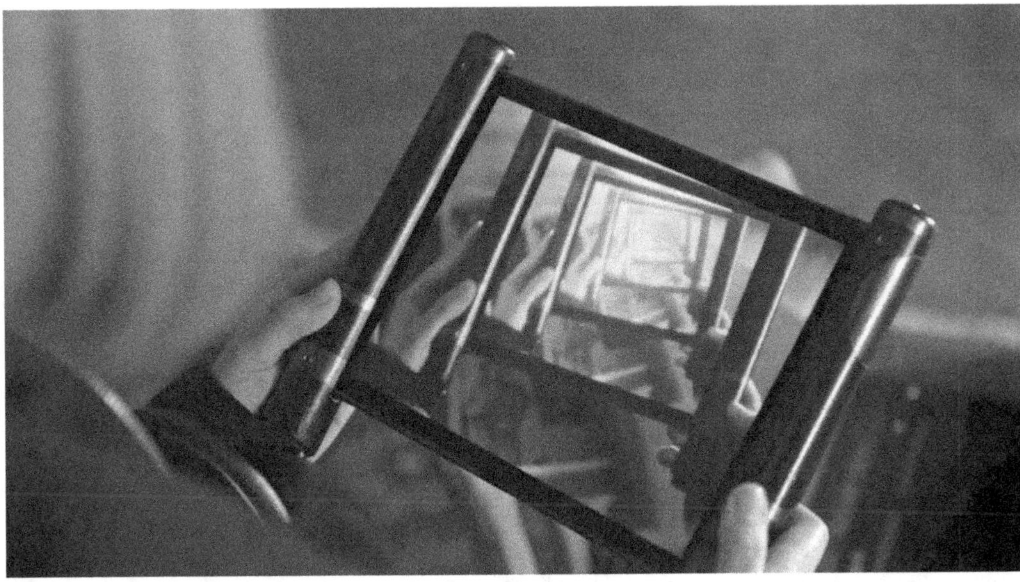

Figure 30: *Mise en abyme:* Rachel discovers that her eye is a camera, constantly monitored by Westmorland.

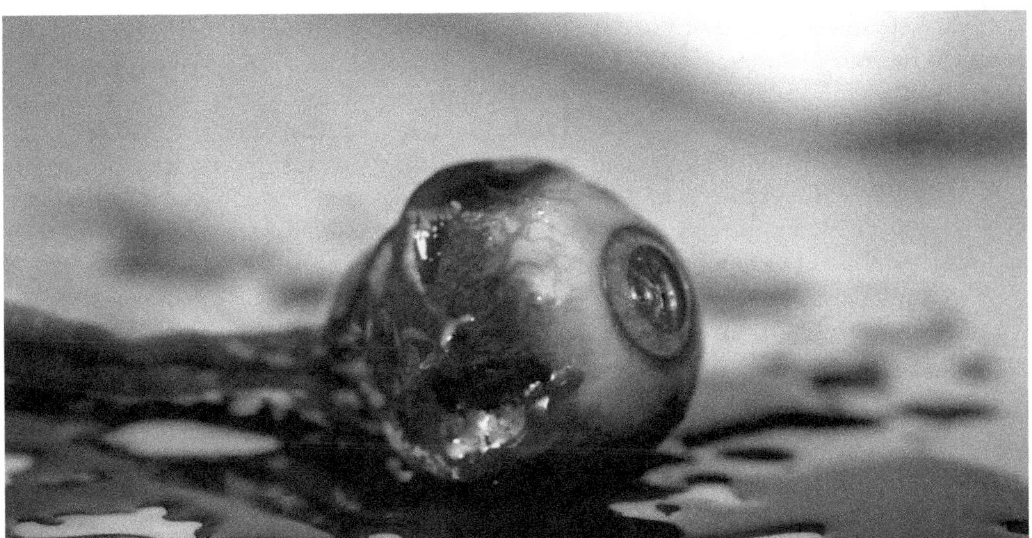

Figure 31: The aftermath of Rachel's 'bloody and redemptive act' in 5.7.

(For all that Maslany focused on the clones' musculature and physicality, the real magic is somehow in the eyes. Maslany's gaze reminds me of those handwarmers full of seed crystals: flip the metal disk, and it resolves into Sarah's desperate intelligence, Alison's pent-up grief, Cosima's openness or Helena's blend of psychopathy and softness.)

I asked her about Rachel. 'I love that Rachel has now been incapacitated at the end of the season', Maslany replies, 'this woman with this perfect veneer, this perfect façade, and it's just been penetrated, and what would that mean for her if she were to survive it?' (Maslany 2014).

Well, now we know. Rachel's final conversion to the clones' side happens when she stares at a screen portraying her own perceptions and realizes they're fatally compromised (Figure 30).

This is in many ways the show's key moment: the moment the clone realizes she's even less than she thought: rather than something for scientists to stare *at*, she's little more than a lens for them to stare through.

I like to imagine that what sent Rachel – who has a very high tolerance for being surveilled – over the edge in this scene is the fact that this new surveillance technology is literally and specifically useless *to her*. Rachel can stand to be looked at, but what P.T. Westmorland sees when he looks at that screen isn't her at all, but a straightforward image of what *she herself sees*. She, herself, has become transparent, uninteresting. Worse still, what *Rachel* sees when she looks at that screen is an unusable infinite regression. She – who has always fancied herself the 'self-aware' clone, the one in on the joke, the one conversant in the studies, the one to whom nothing is opaque – has (at the very moment she was promised emancipation) discovered that she's instead become a technology, a lens that she is by definition unable to use.

It's hard to imagine a richer portrayal of the problems of patriarchy. To be trapped and betrayed by your own perceptions – to be unable to access the thing about yourself that makes you useful: these are expansions of the show's early interest in bodily autonomy and reproductive rights. Rachel's bloody and redemptive act (see Figure 31) suggests that opting out of the system that created you is more than painful. Internalized misogyny is complicated and pervasive, and its distortions aren't easy thing to pluck out. To see clearly – to achieve a mode of perception unclouded by the agar of the petri dish in which you grew – might even demand something perilously close to self-annihilation.

Notes

1 This has always struck me: Rachel's understanding of her institutional status created a longing for some kind of conventional family structure, but it was always fathers, specifically, that appealed to her. Her affect towards Susan Duncan – her first and only 'mother' – never came close to the intensity of feeling the character displayed when Duncan committed suicide (lamenting that she didn't deserve him). Rachel's interactions with her various parental figures always devolve into the latter's expressions of rejection, horror and regret. But she

longs for a father figure who approves of her in ways the other clones don't. Not one – to my knowledge – has ever mentioned a father (Helena never quite seemed to think of Tomas in that way). This is confirmed by the fact that Kendall – phenotypically a woman – turned out to be father and mother both.

2 Alison's ethical code seems to be a compromise between these two extremes – I wrote a little about how her effort to come to terms with Aynsley's death is acting as a proxy for some of the show's murkier moral questions in a column called 'Can *Orphan Black* escape its tangled morality?' for *The Week* (Loofbourow 2017).

3 A review of Rachel's perplexing arc is also available in that column.

4 Religious overtones aside, this is clearly a reference to Dolly the sheep, the first mammal cloned successfully from an adult cell – as was M.K.'s mask, of course.

Appendix: *Orphan Black* Episodes

Season	Episode	Title	Director	Writer	Originally Aired
1	1	'Natural Selection'	John Fawcett	Graeme Manson	30 March 2013
1	2	'Instinct'	John Fawcett	Graeme Manson	6 April 2013
1	3	'Variation under Nature'	David Frazee	Graeme Manson	13 April 2013
1	4	'Effects of External Conditions'	Grant Harvey	Karen Walton	20 April 2013
1	5	'Conditions of Existence'	T.J. Scott	Alex Levine	27 April 2013
1	6	'Variations Under Domestication'	John Fawcett	Will Pascoe	4 May 2013
1	7	'Parts Developed in an Unusual Manner'	Brett Sullivan	Tony Elliott	11 May 2013
1	8	'Entangled Bank'	Ken Girotti	Karen Walton	18 May 2013
1	9	'Unconscious Selection'	T.J. Scott	Alex Levine	25 May 2013
1	10	'Endless Forms Most Beautiful'	John Fawcett	Graeme Manson	1 June 2013

(*Cont.*)

2	1	'Nature Under Constraint and Vexed'	John Fawcett	Graeme Manson	19 April 2014
2	2	'Governed by Sound Reason and True Religion'	John Fawcett	Karen Walton, Graeme Manson	26 April 2014
2	3	'Mingling Its Own Nature with It'	T.J. Scott	Alex Levine	3 May 2014
2	4	'Governed as It Were by Chance'	David Frazee	Russ Cochrane	10 May 2014
2	5	'Ipsa Scientia Potestas Est'	Helen Shaver	Tony Elliott	17 May 2014
2	6	'To Hound Nature in Her Wanderings'	Brett Sullivan	Chris Roberts	24 May 2014
2	7	'Knowledge of Causes and Secret Motion of Things'	Ken Girotti	Aubrey Nealon	31 May 2014
2	8	'Variable and Full of Perturbation'	John Fawcett	Karen Walton	7 June 2014
2	9	'Things Which Have Never Yet Been Done'	T.J. Scott	Alex Levine	14 June 2014
2	10	'By Means Which Have Not Yet Been Tried'	John Fawcett	Graeme Manson	21 June 2014
3	1	'The Weight of This Combination'	David Frazee	Graeme Manson	18 April 2015
3	2	'Transitory Sacrifices of Crisis'	John Fawcett	Aubrey Nealon	25 April 2015
3	3	'Formalized, Complex and Costly'	John Fawcett	Chris Roberts	2 May 2015

3	4	'Newer Elements of Our Defence'	Chris Grismer	Russ Cochrane	9 May 2015
3	5	'Scarred by Many Past Frustrations'	David Frazee	Alex Levine	16 May 2015
3	6	'Certain Agony of the Battlefield'	Helen Shaver	Aubrey Nealon	23 May 2015
3	7	'Community of Dreadful Fear and Hate'	Ken Girotti	Sherry White	30 May 2015
3	8	'Ruthless in Purpose and Insidious in Method'	Aaron Morton	Graeme Manson, Chris Roberts	6 June 2015
3	9	'Insolvent Phantom of Tomorrow'	Vincenzo Natali	Russ Cochrane	13 June 2015
3	10	'History Yet to Be Written'	John Fawcett	Graeme Manson	20 June 2015
4	1	'The Collapse of Nature'	John Fawcett	Graeme Manson	14 April 2016
4	2	'Transgressive Border Crossing'	John Fawcett	Russ Cochrane	21 April 2016
4	3	'The Stigmata of Progress'	Ken Girotti	Aubrey Nealon	28 April 2016
4	4	'From Instinct to Rational Control'	Peter Stebbings	Alex Levine	5 May 2016
4	5	'Human Raw Material'	David Wellington	Kate Melville	12 May 2016
4	6	'The Scandal of Altruism'	Grant Harvey	Chris Roberts	19 May 2016
4	7	'The Antisocialism of Sex'	David Frazee	Nikolijne Troubetzkoy, Graeme Manson	26 May 2016

(*Cont.*)

4	8	'The Redesign of Natural Objects'	Aaron Morton	Peter Mohan	2 June 2016
4	9	'The Mitigation of Competition'	David Frazee	Alex Levine	9 June 2016
4	10	'From Dancing Mice to Psychopaths'	John Fawcett	Graeme Manson	16 June 2016
5	1	'The Few Who Dare'	John Fawcett	Graeme Manson	10 June 2017
5	2	'Clutch of Greed'	John Fawcett	Jeremy Boxen	17 June 2017
5	3	'Beneath Her Heart'	David Wellington	Alex Levine	24 June 2017
5	4	'Let the Children and the Childbearers Toil'	David Wellington	Greg Nelson	1 July 2017
5	5	'Ease for Idle Millionaires'	Helen Shaver	Jenn Engels	8 July 2017
5	6	'Manacled Slim Wrists'	Grant Harvey	David Bezmozgis	15 July 2017
5	7	'Gag or Throttle'	David Frazee	Renée St. Cyr	22 July 2017
5	8	'Guillotines Decide'	Aaron Morton	Aisha Porter-Christie, Graeme Manson	29 July 2017
5	9	'One FetteredSlave'	David Frazee	Alex Levine	5 August 2017
5	10	'To Right the Wrongs of Many'	John Fawcett	Renée St. Cyr, Graeme Manson	12 August 2017

References

Agamben, Giorgio (2000), *Means without End: Notes on Politics* (trans. Vincenzo Binetti and Cesare Casarino), Minneapolis: University of Minnesota Press.

——— (2005), *State of Exception* (trans. Kevin Attell), Chicago: University of Chicago Press.

——— (2011), *Nudities* (trans. David Kishik and Stefan Pedatella), Stanford: Stanford University Press.

Amidi, Amid (2014), 'Andy Serkis does everything, animators do nothing, says Andy Serkis', Cartoon Brew, 8 May, http://www.cartoonbrew.com/motion-capture/andy-serkis-does-everything-animators-do-nothing-says-andy-serkis-98868.html. Accessed 2 June 2017.

Aurthur, Kate (2013), 'Why Patton Oswalt thinks "Orphan Black's" Tatiana Maslany is the "Best Actress Alive"', *BuzzFeed*, 21 June, https://www.buzzfeed.com/kateaurthur/tatiana-maslany-orphan-black-emmys-patton-oswalt. Accessed 30 May 2017.

Austin, John (1975), *How to Do Things with Words*, Oxford: Oxford University Press.

Balzac, Honoré de (1976), 'Théorie de la demarche' ('Theory of the gait'), in Pierre-Georges Castex (ed.), *La Comédie Humaine*, vol. 12, Paris: Gallimard, pp. 259–306.

Barad, Karen (2003), 'Posthumanist performativity: Toward an understanding of how matter comes to matter', *Signs: Journal of Women in Culture and Society*, 28:3, pp. 801–31.

Baron, Cynthia and Carnicke, Sharon Marie (2010), *Reframing Screen Performance*, Michigan: University of Michigan Press.

Baudrillard, Jean (2001), *The Vital Illusion*, New York: Columbia University Press.

——— (2006), *Simulacra and Simulation* (trans. Sheila Faria Glaser), Ann Arbor: University of Michigan Press.

Bazin, André (1992), *Jean Renoir*, New York: Da Capo Press.

BBC America (2012), '*Orphan Black* Teaser – BBC America New Series March 30', YouTube, https://www.youtube.com/watch?v=i-UEa_BIHqU. Accessed 20 May 2017.

——— (2013a), '*Orphan Black* insider: Three clones, one frame', YouTube, https://youtu.be/4rALHr8gV1E. Accessed 15 July 2017.

——— (2013b), '*Orphan Black* clones on clones: Comic-con exclusive making of', YouTube, https://youtu.be/yNfTJCwQ6ms. Accessed 15 July 2017.

——— (2013c), 'Resurrecting Clones & Cosima's Tattoos', YouTube, https://www.youtube.com/watch?v=cgy9wBJn0Tw.

——— (2014a), 'Inside *Orphan Black:* Meet Tatiana's clone double', YouTube, https://youtu.be/9fZLiRTL0-Y. Accessed 15 July 2017.

——— (2014b), 'Making of *Orphan Black*'s 4 Clone Dance Party', YouTube, https://www.youtube.com/watch?v=XE2u_N8g6cs. Accessed 29 May 2018.

——— (2017a), 'OFFICIAL *Orphan Black* Season 5 trailer', YouTube, https://www.youtube.com/watch?v=HFUXSYn7ltw. Accessed 23 May 2017.

——— (2017b), 'Best clone swaps | *Orphan Black* top moments | BBC America', YouTube, https://youtu.be/yizD1FQh-Lo. Accessed 15 July 2017.

——— (2017c), 'Alex Levine: My favorite clone to write', 24 June, http://www.bbcamerica.com/shows/orphan-black/blog/2017/06/alex-levine-my-favorite-clone-to-write

——— (2017d), 'Inside *Orphan Black:* Ease for Idle Millionaires', YouTube, https://www.youtube.com/watch?v=Ykyc7m63kpc Accessed 29 May 2018.

Beard, William (2006), *The Artist as Monster: The Cinema of David Cronenberg*, Toronto: University of Toronto Press.

Bloom, Harold (2013), 'Introduction', in Harold Bloom (ed.), *Aldous Huxley's Brave New World*, Philadelphia: Camden House.

Bode, Lisa (2010), 'No longer themselves? Framing digitally enabled posthumous "performance"', *Cinema Journal*, 4, pp. 46–70.

Bolter, Jay D. and Grusin, Richard (1999), *Remediation: Understanding New Media*, Cambridge: MIT Press.

Bordwell, David and Thompson, Kristin (2012), *Film Art: An Introduction*, 10th ed., New York: McGraw-Hill Education.

Bricken, Rob (2013), '*Orphan Black* reveals how it can film three Tatiana Maslanys all at once', Gizmodo, 24 July, io9.gizmodo.com/orphan-black-reveals-how-it-can-film-three-tatiana-masl-898101511. Accessed 18 March 2018.

Brown, Tracy (2016), 'Tatiana Maslany finally wins an Emmy: Clone Club celebrates', *Los Angeles Times*, 19 September, http://www.latimes.com/entertainment/tv/showtracker/la-et-st-emmys-live-updates-20160918-htmlstory.html. Accessed 2 June 2017.

Bukataman, Scott (2012), *The Poetics of Slumberland: Animated Spirits and the Animating Spirit*, Berkeley: University of California Press.

Butler, Judith (1990), *Gender Trouble: Feminism and the Subversion of Identity*, New York: Routledge.

——— (2016), 'Rethinking vulnerability and resistance', in J. Butler, Z. Gambetti and L. Sabsay (eds), *Vulnerability in Resistance*, Durham: Duke University Press, pp. 12–27.

Čapek, Karel (2004), *R.U.R. (Rossum's Universal Robots)* (trans. C. Novack-Jones), New York: Penguin.

Certeau, Michel de (1984), *The Practice of Everyday Life* (trans. Steven F. Rendall), Berkeley: University of California Press.

Cima, Gay Gibson (1996), *Performing Women: Female Characters, Male Playwrights, and the Modern Stage*, Ithaca: Cornell University Press.

Citton, Yves (2012), *Gestes d'humanités: Anthropologie sauvage de nos expériences esthétiques*, Paris: Armand Colin.

Cohen, Ed (2009), *A Body Worth Defending: Immunity, Biopolitics, and the Apotheosis of the Modern Body*, Durham: Duke University Press.

Dantec, Maurice G. (1999), *Babylon Babies*, Paris: Gallimard.

Deery, June (2013), 'Technology and gender in Aldous Huxley's alternative (?) worlds', in Harold Bloom (ed.), *Aldous Huxley's Brave New World*, Philadelphia: Camden House, pp. 129-46.

Deleuze, Gilles and Guattari, Felix (1987), *A Thousand Plateaus: Capitalism and Schizophrenia*, 2nd ed., Minneapolis: University of Minnesota Press.

Derrida, Jacques (1998), 'Faith and knowledge: The two sources of "religion" at the limits of reason alone', in J. Derrida and G. Vattimo (eds), *Religion*, Cambridge: Polity, pp. 1–78.

——— (2005), *Rogues: Two Essays on Reason*, Stanford: Stanford University Press.

Doyle, Arthur Conan (2005), *The New Annotated Sherlock Holmes* (ed. Leslie Klinger), New York: W. W. Norton & Co.

Edkins, Jenny (2015), *Face Politics*, London and New York: Routledge.

Edwards, Graham (2015), '*Orphan Black* and twinning in the movies', Cinefex Blog, 19 April, http://cinefex.com/blog/orphan-black/. Accessed 30 May 2017.

Edwards, Ronald (2016), *The Edge of Evolution: Animality, Inhumanity, and Doctor Moreau*, New York: Oxford University Press.

Elan, Priya (2014), '*Orphan Black*: The myriad faces of Tatiana Maslany', *The Guardian*, 26 April, https://www.theguardian.com/tv-and-radio/2014/apr/26/tatiana-maslany-orphan-black. Accessed 19 June 2017.

Enelow, Shonni (2015), *Method Acting and Its Discontents: On American Psycho-Drama*, Evanston: Northwestern University Press.

Esposito, Roberto (2006), interviewed by Timothy Campbell (trans. Anna Paparcone), *Diacritics*, 36:2, pp. 49–56.

——— (2008), *Bios: Biopolitics and Philosophy* (trans. Timothy Campbell), Minneapolis: University of Minnesota Press.

——— (2011), *Immunitas: The Protection and Negation of Life* (trans. Zakiya Hanafi), Cambridge: Polity.

——— (2012), *Terms of the Political: Community, Immunity, Biopolitics* (trans. Rhiannon Noel Welch), New York: Fordham University Press.

Fawcett, John (2014), interviewed by Lili Loofbourow, Toronto, 3 December.

——— (2015), interviewed by Alicia Lutes, Nerdist, https://nerdist.com/orphan-black-co-creator-john-fawcett-talks-crazy-season-3/. Accessed 29 May 2015.

Firchow, Peter Edgerly (1984), *The End of Utopia: A Study of Aldous Huxley's Brave New World*, Lewisburg: Bucknell University Press.

Follows, Stephen (2014), 'What percentage of a film crew is female?', StephenFollows.com, 22 July, http://stephenfollows.com/gender-of-film-crews/. Accessed 2 December 2017.

Foucault, Michel (1978), *The History of Sexuality, Vol. 1: An Introduction* (trans. Robert Hurley), New York: Vintage.

——— (1995), *Discipline and Punish: The Birth of the Prison* (trans. Alan Sheridan), New York: Vintage Books.

——— (2003), *Society Must Be Defended: Lectures at the Collège de France, 1975–76* (trans. David Macey), New York: Picador.

Freud, Sigmund ([1908]1959), 'Family romances', in James Strachey (ed.) *Standard Edition*, vol. IX, London: Hogarth Press, pp. 235–41.

——— (2003), *The Uncanny* (trans. David McLintock), New York: Penguin.

Fringe (2008–13, USA: Fox).

Gallagher-Ross, Jacob (2015), 'Mediating the method', *Theatre Survey*, 56:3, pp. 291–313.

Goh, Irving (2015), *The Reject: Community, Politics, and Religion after the Subject*, New York: Fordham University Press.

Grobe, Christopher (2016a), 'Every nerve keyed up: "Telegraph plays" and networked performance, 1850–1900', *Theater*, 46:2, pp. 7–33.

——— (2016b), 'Why it's "easier to act with a telephone than a man"', *Theatre Survey*, 57:2, pp. 175–99.

Hamner, Everett (2017), 'Epigenetic television: The penetrating love of *Orphan Black*', *Los Angeles Review of Books*, 9 June, https://lareviewofbooks.org/article/epigenetic-television-the-penetrating-love-of-orphan-black/. Accessed 10 June 2017.

Haraway, Donna (1991), 'A cyborg manifesto: Science, technology, and socialist-feminism in the late twentieth century', in *Simians, Cyborgs, and Women: The Reinvention of Nature*, New York: Routledge, pp. 149–81.

——— (2007), *When Species Meet*, Minneapolis: University of Minnesota Press.

Hewett, Richard (2015), 'The changing determinants of UK television acting', *Critical Studies in Television*, 10:1, pp. 73–90.

Hiatt, Brian (2015), 'Tatiana Maslany on that Emmys snub and playing nine roles on *Orphan Black*', *Rolling Stone*, 27 April, http://www.rollingstone.com/tv/features/tatiana-maslany-on-that-emmys-snub-and-playing-nine-roles-on-orphan-black-20150427. Accessed 2 June 2017.

Hillegas, Mark R. (1967), *The Future as Nightmare: H.G. Wells and the Anti-Utopians*, New York: Oxford University Press.

Huang, S.L. (2014), 'Watching *Orphan Black* as a faceblind person', goodreads, http://www.goodreads.com/author_blog_posts/6609865-watching-orphan-black-as-a-faceblind-person. Accessed 18 March 2018.

Humann, Heather D. (2016), 'BBC America's *Orphan Black*: Clones and cultural anxieties at the intersection of cult and quality television', *Americana: E-Journal of American Studies in Hungary*, 12:2, http://americanaejournal.hu/vol12no2/humann. Accessed 12 May 2017.

Huxley, Aldous (1932), *Brave New World*, Toronto: HarperCollins.

Huxley, Thomas Henry ([1894] 1989), *Evolution and Ethics*, Princeton: Princeton University Press.

Irwin, John T. (1994), *The Mystery to a Solution: Poe, Borges, and the Analytic Detective Story*, Baltimore: The Johns Hopkins University Press.

Jenkins, Henry (2006), *Fans, Bloggers, and Gamers: Media Consumers in a Digital Age*, New York: NYU Press.

Jentsch, Ernst (1993), 'On the psychology of the uncanny' (trans. Roy Sellars), *Angelaki*, 2:1, pp. 7–16.

Kapur, Isabella (2013), 'Clones are people too: The science and science fiction of BBC America's *Orphan Black*', The Mary Sue, 15 August, https://www.themarysue.com/orphan-black-science-and-fiction/. Accessed 13 March, 2018.

Keefe, Patrick R. (2016), 'The detectives who never forget a face', *The New Yorker*, 22 August, http://www.newyorker.com/magazine/2016/08/22/londons-super-recognizer-police-force. Accessed 30 May 2017.

Keller, Evelyn F. (2000), *The Century of the Gene*, Cambridge: Harvard University Press.

Kirby, Michael (1972), 'On acting and not-acting', *The Drama Review: TDR*, 16:1, pp. 3–15.

Kittler, Friedrich (1999), *Gramophone, Film, Typewriter* (trans. Geoffrey Winthrop-Young and Michael Wutz), Stanford: Stanford University Press.

Kleist, Heinrich von (1972), 'On the marionette theatre', *The Drama Review*, 16:3, pp. 22–26.

Koblin, John (2017), 'Forget ratings: *Orphan Black* had the #CloneClub', *New York Times*, 10 August, https://www.nytimes.com/2017/08/10/business/media/orphan-black-cloneclub.html. Accessed 15 December 2017.

Kuleshov, Lev V. (1974), *Kuleshov on Film: Writings* (trans. Ronald Levaco), Berkeley: University of California Press.

Lacan, Jacques (1993), *The Seminar of Jacques Lacan, Book III: The Psychoses, 1955–1956* (ed. Jacques-Alain Miller, trans. Russell Grigg), New York: Norton.

——— (2006), 'The instance of the letter in the unconscious, or reason since Freud', in *Écrits: The First Complete Edition in English* (trans. Bruce Fink), New York: Norton, pp. 412–44.

LaValley, Albert J. (1985), 'Traditions of trickery: The role of special effects in the science fiction film', in George Slusser and Eric S. Rabkin, *Shadows of the Magic Lamp: Fantasy and Science Fiction in Film*, Carbondale and Edwardsville: Southern Illinois University Press, 141–58.

Lepore, Jill (2015), 'The history lurking behind *Orphan Black*', *The New Yorker*, 16 April, http://www.newyorker.com/culture/cultural-comment/the-history-lurking-behind-orphan-black. Accessed 30 May 2017.

Levin, Ira (1976), *The Boys from Brazil*, New York: Random House.

Levine, Alex (2014), interviewed by Lili Loofbourow, Toronto, 1 December.

Littlefield, Melissa M. and Johnson, Jenell M. (eds) (2012), *The Neuroscientific Turn: Transdisciplinarity in the Age of the Brain*, Ann Arbor: The University of Michigan Press.

Locker, Melissa (2015), 'Tatiana Maslany explains how she played a clone playing a clone on *Orphan Black*', *Vanity Fair*, 20 April, http://www.vanityfair.com/hollywood/2015/04/tatiana-maslany-orphan-black-premiere-interview. Accessed 1 June 2017.

Loofbourow, Lili (2015), 'The many faces of Tatiana Maslany', *New York Times Magazine*, 2 April, https://www.nytimes.com/2015/04/05/magazine/the-many-faces-of-tatiana-maslany.html. Accessed 13 March 2018.

——— (2017), 'Can *Orphan Black* escape its tangled morality?', *The Week*, 26 June, http://theweek.com/articles/707219/orphan-black-escape-tangled-morality. Accessed 13 March 2018.

Lutes, Alicia (2015), '*Orphan Black* co-creator John Fawcett talks "crazy" season 3', Nerdist, 21 January, https://nerdist.com/orphan-black-co-creator-john-fawcett-talks-crazy-season-3/ Accessed 29 May 2015.

Lynch, Stephen (2014), interviewed by Lili Loofbourow, Toronto, 1 December.

Malabou, Catherine (2016), 'One life only: Biological resistance, political resistance' (trans. Carolyn Shread), *Critical Inquiry*, 42, pp. 429–38.

Manning, Erin (2013), *Always More Than One: Individuation's Dance*, Durham: Duke University Press.

Manovich, Lev (1995), 'The paradoxes of digital photography', http://manovich.net/content/04-projects/004-paradoxes-of-digital-photography/02_article_1994.pdf. Accessed 13 March 2018.

——— (1996), 'Cinema and digital media', http://manovich.net/content/04-projects/008-cinema-and-digital-media/06_article_1995.pdf. Accessed 13 March 2018.

Manson, Graeme (2014), interviewed by Lili Loofbourow, Toronto, 2 December.

Markham, A. and Buchanan, E. (2012), *Ethical Decision-Making and Internet Research: Recommendations from the AoIR Ethics Working Committee (Version 2.0)*, Association of Internet Researchers, https://aoir.org/reports/ethics2.pdf. Accessed 30 May 2017.

Marks, Leo (1998), *Between Silk and Cyanide: A Codemaker's War, 1941–45*, London: Harper Collins.

Maslany, Tatiana (2014), interviewed by Lili Loofbourow, Toronto, 2 December.

Mathijs, Ernest (2013), *John Fawcett's* Ginger Snaps, Toronto: University of Toronto Press.

May, Emmeline (2015), 'So tell me who are you?', ROCKSTAR·DINOSAUR·PIRATE·PRINCESS, 6 April, http://rockstardinosaurpirateprincess.com/2015/04/06/so-tell-me-who-are-you/. Accessed 18 March 2018.

Mbembe, Achille (2013), 'Necropolitics', in Timothy Campbell (ed.), *Biopolitics: A Reader*, Durham: Duke University Press, pp. 161–92.

McConnell, Frank (1981), *The Science Fiction of H.G. Wells*, New York: Oxford University Press.

McGeough, Maria (2015), 'Technodolly and the art of authentic cloning in *Orphan Black*', *Motherboard (Vice)*, 22 April, http://motherboard.vice.com/read/technodollies-and-the-art-of-authentic-cloning-in-orphan-black. Accessed 18 September 2015.

McGowan, Todd (2007), *The Impossible David Lynch*, New York: Columbia University Press.

McLeod, Randall (2011), 'Sortes virgilianae', *English Institute Conference*, Harvard University, 10 September.

McLuhan, Marshall (1964), *Understanding Media: The Extensions of Man*, New York: Signet Books.

McMillan, Graeme (2014), 'The cult of *Orphan Black*', *Time*, 21 April, http://time.com/70179/orphan-black-fan-following-reception/. Accessed 4 February 2016.

McNamara, Mary (2014), '*Orphan Black* turns out to be many times better than expected', *Los Angeles Times*, 18 April, http://www.latimes.com/entertainment/tv/showtracker/la-et-st-orphan-black-review-20140419-story.html#axzz2zHW4bpXc. Accessed 13 March 2018.

McTeague, James H. (1993), *Before Stanislavsky: American Professional Acting Schools and Acting Theory, 1875–1925*, Metuchen: Scarecrow Press.

Mele, Rick (2013), '*Orphan Black:* 10 things you need to know', *Huffington Post Canada*, 28 March, http://www.huffingtonpost.ca/2013/03/28/orphan-black-revew-set-visit_n_2972315.html. Accessed 13 March 2018.

Metz, Christian (1977), '*Trucage* and the film' (trans. Françoise Meltzer), *Critical Inquiry*, 3:4, pp. 657–75.

Miller, Julie (2015), 'Here's how *Orphan Black* transforms Tatiana Maslany into a cast of clones', *Vanity Fair*, 20 April, http://www.vanityfair.com/hollywood/2015/04/orphan-black-clones. Accessed 30 May 2017.

Morris, Aaron (2015), 'Tatiana Maslany creates playlists for each of her *Orphan Black* characters', *Noisey (Vice)*, 14 April, https://noisey.vice.com/en_ca/article/orphan-black-tatiana-maslany-interview-playlist-spotify-2015. Accessed 30 May 2017.

Murphy, Megan (2016), '*Orphan Black* returns to form', *New Republic*, 14 April, https://newrepublic.com/article/132666/orphan-black-returns-form. Accessed 13 March 2018.

Naremore, James (1988), *Acting in the Cinema*, Berkeley: University of California Press.

Nemetz, Dave (2014), 'How *Orphan Black*'s hair and makeup wizards split one actress into five (very distinct) clones', Yahoo! TV, 15 May, http://tv.yahoo.com/blogs/tv-news/orphan-black-hair-makeup-secrets-tatiana-maslany-223253089.html. Accessed 30 May 2017.

Neroni, Hilary (2015), *The Subject of Torture: Psychoanalysis & Biopolitics in Television and Film*, New York: Columbia University Press.

Ngai, Sianne (2004), *Ugly Feelings*, Cambridge: Harvard University Press.

North, Daniel (2010), 'Evidence of things not quite seen: *Cloverfield*'s obstructed spectacle', *Film & History*, 40:1, pp. 75–92.

Nussbaum, Emily (2014), 'Cheaper by the dozen', *The New Yorker*, 28 April, http://www.newyorker.com/magazine/2014/04/28/cheaper-by-the-dozen. Accessed 19 June 2017.

@orphanblack (2017), 'What clone swap do you want to see for the final trip? #OrphanBlack', Twitter, 24 April, https://twitter.com/orphanblack/status/856563836836495360. Accessed 2 June 2017.

Pacific Motion Control (n.d.), 'Technodolly', http://pacificmotion.net/programmable-technodolly/. Accessed 18 September 2015.

Pence, Gregory E. (2016), *What We Talk About When We Talk About Clone Club: Bioethics and Philosophy in* Orphan Black, Dallas: BenBella Books.

Pitches, Jonathan (2006), *Science and the Stanislavsky Tradition of Acting*, New York: Routledge.

Poe, Edgar A. (1966), 'The Gold-Bug', in *The Complete Stories and Poems of Edgar Allan Poe*, New York: Doubleday, pp. 70–96.

Prudom, Laura (2015), '*Orphan Black* Season 3: Tatiana Maslany on male clones, "strong female characters"', *Variety*, 17 April, http://variety.com/2015/tv/news/orphan-black-tatiana-maslany-season-3-1201474698/. Accessed 30 May 2017.

——— (2017), '*Orphan Black* has changed the game for TV visual effects, and half of them are invisible', Mashable, 9 June, http://mashable.com/2017/06/09/orphan-black-season-5-visual-effects-clones-geoff-scott/. Accessed 15 June 2017.

Racine, Eric and Zimmerman, Emma (2012), 'Pragmatic neuroethics and neuroscience's potential to radically change ethics', in Melissa M. Littlefield and Jenell M. Johnson (eds), *The Neuroscientific Turn: Transdisciplinarity in the Age of the Brain*, Ann Arbor: The University of Michigan Press, pp. 135–51.

Rank, Otto ([1925] 1971), *The Double: A Psychoanalytic Study* (trans. and ed. Harry Tucker, Jr.), Chapel Hill: University of North Carolina Press.

Recht, Hannah (2016), *Orphan Black: Data Analysis*, http://hrecht.github.io/orphanblack-update/. Accessed 2 June 2017.

Reddit (2015), 'Comment: Am I the only one confused on Tatiana Maslany's face?', www.reddit.com/r/orphanblack/comments/2rjle6/am_i_the_only_one_confused_on_tatiana_maslany_face/. Accessed 18 March 2018.

——— (2016), 'Comment: DAE forget that the characters are played by the same person?', www.reddit.com/r/orphanblack/comments/36ck8r/dae_forget_that_the_characters_are_played_by_the/. Accessed 3 March 2018.

Redfield, Marc (1996), *Phantom Formations: Aesthetic Ideology and the* Bildungsroman, Ithaca: Cornell University Press.

Rickitt, Richard (2000), *Special Effects: The History and Technique*, New York: Billboard Books.

Roach, Joseph R. (1993), *The Player's Passion: Studies in the Science of Acting*, Ann Arbor: University of Michigan Press.

Robey, Tim (2014), 'Does Andy Serkis's motion capture acting deserve an Oscar?', *Telegraph*, 8 November, http://www.telegraph.co.uk/culture/film/oscars/11217293/Does-Andy-Serkiss-motion-capture-acting-deserve-an-Oscar.html. Accessed 2 June 2017.

Romanello, Linda (2015), 'VFX for TV: *Orphan Black*', *Post*, 1 April, http://www.postmagazine.com/Publications/Post-Magazine/2015/April-1-2015/VFX-For-TV-Orphan-Black.aspx. Accessed 18 September 2015.

Rosenheim, Shawn J. (1989), 'The king of "Secret Readers": Edgar Poe, cryptography, and the origins of the detective story', *ELH*, 56:2, pp. 375–400.

——— (1997), *The Cryptographic Imagination: Secret Writing from Edgar Poe to the Internet*, Baltimore: The Johns Hopkins University Press.

Ryan, Maureen (2013), '*Orphan Black:* A sexy identity thriller that'll keep you guessing', *Huffington Post Canada*, 29 May, http://www.huffingtonpost.com/maureen-ryan/orphan-black_b_2974964.html. Accessed 13 March 2018.

Sacks, Oliver (2010), 'Face-blind', *The New Yorker*, 30 August, https://www.newyorker.com/magazine/2010/08/30/face-blind. Accessed 29 May 2018.

Salvato, Nick (2015), 'Big glove: Televisual dissociation and embodied performance', *Criticism*, 57, pp. 91–107.

Saraiya, Sonia (2014), '*Orphan Black* clones itself, embracing the comedy', The AV Club (The Onion), 20 June, http://www.avclub.com/review/orphan-black-clones-itself-embracing-comedy-206048. Accessed 13 March 2018.

Shacklock, Zoë (2016), 'Two of a kind: Revaluing the work of acting doubles in *Orphan Black*', *Journal of Film and Video*, 68:3, pp. 69–82.

Sheets-Johnstone, Maxine (2011), *The Primacy of Movement*, 2nd ed., Amsterdam and Philadelphia: John Benjamins.

Silvio, Teri (2010), 'Animation: The new performance?', *Journal of Linguistic Anthropology*, 20, pp. 422–38.

Singh, Simon (1999), *The Code Book: The Science of Secrecy from Ancient Egypt to Quantum Cryptography*, New York: Doubleday.

Spielberg, Steven (1993), *Jurassic Park*, USA: Universal Studios.

Stewart, Garrett (1985), 'The "videology" of science fiction', in George Slusser and Eric S. Rabkin (eds), *Shadows of the Magic Lamp: Fantasy and Science Fiction in Film*, Carbondale and Edwardsville: Southern Illinois University Press, 159–207.

Stutsman, Staci (2016), 'The unruly clones: Tatiana Maslany's melodramatic masquerades in *Orphan Black*', *Journal of Film and Video*, 68:3, pp. 83–103.

Tait, R. Colin (2014), '*Orphan Black* and the performance text of attractions', *In Media Res*, 7 April, http://mediacommons.futureofthebook.org/imr/2014/04/07/orphan-black-and-performance-text-attractions. Accessed 3 June 2017.

Truong, Jean-Michel ([1989] 2015), *Reproduction Interdite*, rev.ed., Paris: Gallimard.

Väliaho, Pasi (2014), *The Biopolitical Image: Image, Power, and the Neoliberal Brain*, Cambridge, MA: MIT Press.

Watson, James (2005), 'How we discovered DNA,' TED, https://www.ted.com/talks/james_watson_on_how_he_discovered_dna/transcript?language=en. Accessed 1 June 2017.

Wells, H.G. ([1896] 2005), *The Island of Doctor Moreau*, New York: Signet.

Wilcox, Ella Wheeler (1912), 'I protest', *The Lather*, 12:9, p. 32.

Wilhelm, Kate ([1976] 1998), *Where Late the Sweet Birds Sang*, New York: Tom Doherty Associates.

Williams, Alan (1992), *Republic of Images: A History of French Filmmaking*, Cambridge, MA: Harvard University Press.

Woerner, Meredith (2014), 'Andy Serkis built a new world for *Dawn of the Planet of the Apes*', io9, 28 March, http://io9.gizmodo.com/andy-serkis-reveals-the-new-ape-world-in-dawn-of-the-pl-1553706020. Accessed 2 June 2017.

Wolf, Mark J. P. (2003), 'The technological construction of performance', *Convergence: The International Journal of Research into New Media Technologies*, 9:4, pp. 48–59.

Wolfinger, Emily (2016), '"But it's already public, right?": The ethics of using online data', *Data Driven Journalism*, 25 November, http://datadrivenjournalism.net/news_and_analysis/but_its_already_public_right_the_ethics_of_using_online_data. Accessed 30 May 2017.

Yahr, Emily (2014), '*Orphan Black* and Tatiana Maslany got snubbed by the Emmy Awards: Here's why that's not shocking', *Washington Post*, 10 July, https://www.washingtonpost.com/news/arts-and-entertainment/wp/2014/07/10/orphan-black-and-tatiana-maslany-got-snubbed-by-the-emmy-awards-heres-why-thats-not-shocking/. Accessed 2 June 2017.

Zimmerman, Danielle (2015), 'Our 5 favorite *Orphan Black* clone swaps (so far)', Hypable, 18 April, http://www.hypable.com/best-orphan-black-clone-swaps/. Accessed 30 May 2017.

Žižek, Slavoj (1991), *Looking Awry: An Introduction to Jacques Lacan through Popular Culture*, Cambridge, MA: MIT Press.

——— (2008), *Violence: Six Sideways Reflections*, New York: Picador.

Notes on Contributors

David F. Bell is professor of romance studies at Duke University. His research interests include nineteenth-century French literature, science studies, philosophy and literature, and the intersections between neuroscience and cognitive studies. He is the author of three books, *Real Time: Accelerating Narrative from Balzac to Zola* (University of Illinois Press, 2004), *Circumstances: Chance in the Literary Text* (University of Nebraska Press, 1993) and *Models of Power: Politics and Economics in Zola's* Rougon-Macquart (University of Nebraska Press, 1988), as well as numerous articles on the nineteenth-century novel and on contemporary French philosophy.

Andrea Goulet is professor and graduate chair of French and francophone studies at the University of Pennsylvania. She is the author of *Optiques: The Science of the Eye and the Birth of Modern French Fiction* (Penn, 2006) and *Legacies of the Rue Morgue: Space and Science in French Crime Fiction* (Penn, 2016), as well as numerous articles on nineteenth- and twentieth-century French fiction, critical theory, science and literature, detective fiction and the *nouveau roman.* She has co-edited special journal issues on 'Visual Culture' (*Contemporary French Civilization*) and 'Crime Fictions' (*Yale French Studies*) and is currently co-chair of the Nineteenth-Century French Studies (NCFS) Association.

Christopher Grobe is associate professor of English at Amherst College and the author of *The Art of Confession: The Performance of Self from Robert Lowell to Reality TV* (NYU, 2017). His work on *Orphan Black* collected here is part of a second book project, on robots and realist acting. His other essays on literature, performance and media culture can be found in *PMLA*, *NLH*, *Theater* and *Theatre Survey*, as well as in several edited collections.

Lili Loofbourow is the staff culture critic at *The Week* and a special correspondent for the *Los Angeles Review of Books*. Her interests include popular culture, Milton, early modern natural philosophy, cognitive science and Chile. She won the 2015 Staige D. Blackford Prize for Nonfiction, and is currently at work on two books.

Hilary Neroni is professor of film and television studies at the University of Vermont. She is the author of *The Violent Woman: Femininity, Narrative, and Violence in Contemporary*

American Cinema (SUNY, 2005), *The Subject of Torture: Psychoanalysis and Biopolitics in Television and Film* (Columbia University Press, 2015) and *Feminist Film Theory and Cléo from 5 to 7* (Bloomsbury, 2016).

Sharrona Pearl is assistant professor of communication at the Annenberg School for Communication at the University of Pennsylvania, and core faculty in the program on gender, sexuality and women's studies. A historian and theorist of the face and body, Pearl has published widely on images and ethics, Victorian visual culture, the history of medicine and religion and media. She is the author of *Face/On: Face Transplants and the Ethics of the Other* (University of Chicago Press, 2017), *About Faces: Physiognomy in Nineteenth-Century Britain* (Harvard University Press, 2010) and editor of *Images, Ethics, Technology* (Routledge, 2015).

Simon Porzak is a co-director in the undergraduate writing program at Columbia University, where he manages pilot programmes that teach reading and writing in collaboration both with the university's Institute for Data Sciences and Department of Music. His research investigates the intersections of science and aesthetics, in particular computer simulations of thought, biomechanics, opera, *fin-de-siècle* literature, affect, sexuality and video games. His essays have appeared in *Diacritics, Modernism/modernity, Symploke, Qui Parle, Nineteenth-Century French Studies, Unwinnable: Life with Culture* and *Opera Quarterly*. He is currently preparing two book-length manuscripts for publication: *Anima Automata: Feeling Technology and the Olympian Art of Song*, on the artificial emotional intelligence of singing dolls from Les Contes d'Hoffmann to Kylie Minogue; and *Against 'Against Nature': Decadent Darwinisms*, on the evolutionary biology of Huysmans, Rachilde, Proust, Lovecraft and Ballard.

Robert A. Rushing is professor of Italian and comparative literature at the University of Illinois, Urbana-Champaign, where he holds affiliate appointments with media and cinema studies and the Unit for Criticism and Interpretive Theory. Professor Rushing's research interests include film and television studies; critical theory; popular culture; comparative studies and genre. He is the author of *Resisting Arrest: Detective Fiction and Popular Culture* (Other Press, 2007), *Descended from Hercules: Biopolitics and the Muscled Male Body on Screen* (Indiana University Press, 2016; winner of the American Association for Italian Studies' best film/media book prize) and co-editor of *Mad Men, Mad World: Sex, Politics, Style and the 1960s* (Duke University Press, 2013).

John C. Stout is an associate professor of French at McMaster University. His research focuses mainly on twentieth-century French poetry. He has also published articles on contemporary American and Canadian experimental poetry. His book *L'Énigme-poésie: Entretiens avec 21 poètes françaises* appeared with Rodopi in 2010. In 2018, he will publish the book *Objects Observed: The Poetry of Things in Twentieth-Century France and America* (University of Toronto Press).

Jessica Tanner is assistant professor of French and francophone studies at the University of North Carolina at Chapel Hill, where she specializes in nineteenth-century French literature and culture and critical theory. Her research interests include space and place; sex and sexuality; biopolitics; technology; environmental humanities and narrative form. She is the author of *Mapping Prostitution: Sex, Space, and the Novel in Nineteenth-Century France* (book manuscript in progress), and the recipient of the 2015 Larry Schehr Memorial Award for best junior faculty essay at the Nineteenth-Century French Studies Colloquium.

Index

Fictional characters are found alphabetically by their first name (e.g., Alison Hendrix), while real persons are to be found alphabetically by their last name (e.g., Hendrix, Jimi).

A

Abrams, J.J., 59, 60, 61
Agamben, Giorgio, 11, 36, 37, 114, 116
Alexandre, Kathryn, 4, 15, 45, 46, 83, 142
Alias, 59, 60
Alison Hendrix, 3, 5, 6, 9, 10, 11, 13, 20n.3, 29, 30, *33*, 35, 36, 43, 44, 53, 54, 60, 62, 81, 83, 84, 85, 86, 97, 98, 99, 107, 108, 119, *121*, 122, 124, 138, 141, 143, 151, 159, 161, 171, 172, 180, 181, 183, 189, 190, 196, 198, 200, 201n.2
Amelia, 13, 189, 191
alphanumeric encoding, 165n.8
animation, 16, 49, 61, 65, 66, 69, 71, 73, 74, 75
Attenborough, Richard, 61, *70–71*
Association of Internet Research (AoIR), 90–91n.3
Aynsley Norris, 5, 201n.2

B

Babylon A.D. (film), 173
Babylon Babies, 20, 172, 176–177, 181. *See also* French sci-fi cloning fiction
Bailey Downs, 13, 21n.4, 107, 109n.3
Balzac, Honoré de, 37n.8
Barad, Karen, 47, 48, 55
Baron, Cynthia, 41, 50
Battlestar Galactica, 66, 123
Baudrillard, Jean, 67, 68, 71, 72, 73
Bazin, André, 31–32, 37n.5
BBC America, 43, 50, 59, 91n.7, 144
Bell, David F., 4, 14, 15, 55n.4
Beth Childs, 3, 4, 5, 6, 18, 43, 44, 48, 60, 62, 76, 82, 83, 84, 86, 89, 97, 103, 121, 129, 132, 138, 141, 144, 159, 180
Bildungsroman, 129, 133
biopolitics, 164n.1, 173, 178; cultural representations of 174–176
Bode, Lisa, 50, 52
Bolter, David, 65
Bonnie Johansson, 17, 98
Bordwell, David, 37n.4
The Boys from Brazil (novel), 17, 100, 105–106
Brave New World, 17, 100, 103–105, 106, 108
BrightBorn, 7, 13, 123, 133, 135, 137, 148, 162, 179, 189
Bruun, Kristian, *9*, *10*
Buchanan, Elizabeth, 91n.3
Bukataman, Scott, 73
Butler, Judith, 11, 139

C

caching, 14, 27–28, 29, 30, 31, 32, 36, 37, 37n.2
Campbell, Timothy, 147
Capek, Čarel, 48
capitalism, 123, 124, 154, 177
Carnicke, Sharon Marie, 41, 50
Castor clones, 8, 16, 17, 48, 98, 101, 103, 105, 106, 108, 132, 136, 137, 138, 140, 150,

161, 163, 170, 178, 189, 190. *See also* Project Castor
The Century of the Gene, 117
Certeau, Michel de, 136, 139
Chevalier, Ferdinand, 129
Cima, Gay, 52–53
Cixous, Hélène Cixous, 179
Clarke, Arthur C., 66
Clément, Catherine, 179, 181
cloning, 5, 12, 13, 15, 16, 17, 18, 20, 48, 63, 67–68, 69, 72, 73, 74, 76, 105, 106, 123, 124, 124n.1, 137, 170, 173, 177, 178
Cloverfield (film), 60
Cohen, Ed, 130
Conan Doyle, Arthur. *See* Doyle, Arthur Conan
Continuum, 12, 164n.1
Cosima Niehaus, 3, 4, 5, 6, 11, 12, 13, 19, 20n.2, 20n.3, 29, 30, 43, 44, 54, 55, 60, 65, 66, 74, *75*, 81, 82, 83, 84, 85, 86, 97, 99, 101, 102, 118, 119, *121*, 122, 123, 124, 136, 137, 138, 139, 141, 142, 143, 144, 149, 151, 152, 153, 154, 155, 156, 159, 161, 162, 169, 170, 171, 172, 175, 180, 181, 183, 187, 190, 194, 196, 198, 200. *See also* Herter, Cosima
Crash (novel), 67–68
Crick, Francis, 115
cryptography, 164n.2
CSI (TV series), 66, 114
Curry, Samuel Silas, 47
Cyborg (painting), *182*
cyborgs, 17, 41, 46, 47–48, 50, 55, 182, 183

D

Dantec, Maurice, 20, 172, 173, 176, 181
Dark Matter, 12
Darwin, Charles, 3, 7, 14, 68, 100, 102, 115, 129, 149, 159, 178, 181, 182
Davis, Bette, 27, 196
Deery, June, 104
Deleuze, Gilles, 16, 80, 82, 90, 91n.8, 171, 176, 177, 179, 181
Delphine Cormier, 19, 65, 66, 137, 144, 149, 152, 153, 155, 161, 162, 169, 170
Derrida, Jacques, 130, 131, 132, 179, 180
Desperate Housewives, 20n.3, 52
Dexter, 20n.3
Dick, Philip K., 176
DNA, 7, 17–18, 69–73, 150, 165n.7
Donnie Hendrix, 4, *9*, *10*, 20–21n.4, 43, 98, 99, 138, 161, 180, 190
The Double: A Psychoanalytic Study, 64, 108n.1
Doyle, Arthur Conan, 17, 19, 129, 149, 151, 152, 155, 156, 157, 161, 162, 163, 164n.3; "The Adventure of the Dancing Men" 19; "The Adventure of the Speckled Band" 164n.5
Dr Aldous Leekie, 4, 6, 7, 17, 65, 66, 98, 99, 102, 118, 133, 160, 162, 172, 188
Dr Nealon, 102, 169
Dr Virginia Coady, 17, 98, 102, 133, 136, 137, 170, 181, 189
Dyad Institute, 7, 60, *65*, 97, 101, 103, 108, 113, 119, 123, 124, 125n.5, 133, 134, 135, 136, 139, 143, 153, 155, 159, 169, 170, 175, 179, 180, 187, 195

E

Edkins, Jenny, 80, 87, 88, 91n.8
Eisenstein, Sergei, 31, 37n.4
Enelow, Shonni, 55n.2
Esposito, Roberto, 18, 130, 131, 132, 133, 135, 136, 140, 143
Ethan Duncan, 99, 101, 152, 154, 155, *156*, 159, 162–163, 171, 172, 188
eugenics, 100–101
Evie Cho, 102, 135, 137, 148, 189
Evolution and Ethics, 100, 102, 109n.2

F

Facebook, 87, 88, 91n.3
facial blindness. *See* prosopagnosia
Fawcett, John, 17, 28, 29, 51–52, 100, 106–107, 108, 196, 197

Felix, 4, 10, 12, 21n.4, 35, 44, 76, 80, 89, 91n.15, 98, 118, 136, 137–138, 144, 161, 170, 180, 181, 190, 191
female subjectivity, 119–121, 122
feminist theory, 8, 47, 49, 55, 59, 64, 144, 162, 169, 172, 182
Ferdinand, 134, 143, 189
Follows, Stephen, 56n.5
Foucault, Michel, 7, 11, 116, 131, 148
French sci-fi cloning fiction, 17, 173. *See also* *Babylon Babies* and *La Reproduction Interdite*
Freud, Sigmund, 17, 64, 66, 98–99, 122
Frewer, Matt, 118
Fringe (TV series), 59, 60

G
Gallagher-Ross, Jacob, 55n.2–3
gestural movement, 33–37, 37n.8
Gilbert, Walter, 115
Ginger Snaps (film), 17, 100, 107, 108, 109n.3
Goh, Irving, 20, 173, 177, 179–180, 181
Goulet, Andrea, 8, 17, 19, 156, 187
Gracie Johanssen, 19, 98, 169, 170, 194
Grobe, Christopher, 4, 5, 8, 15, 30, 47
Grusin, Richard, 65
Guattari, Pierre-Félix, 16, 80, 82, 90, 177, 181

H
Hamner, Everett, 140
Haraway, Donna, 8, 11, 15, 18, 46, 47, 48, 49, 51, 55, 130, 140, 143, 144, 176, 182, 183
Heda Gabler, 53–54
Heddatron, 53
Heidegger, Martin, 68
Helena, 3, 4, 6, 12, 13, 19, 20n.3, 29, 35, 36, 44, 53, 54, 55, 62, 64, 81, 82, 83, 84, 86, 97, 98, 99, 103, 106, 108, 119, *121*, 122, 124, 130, 132, 137, 140, *141*, 142, 143, 144, 149, 150, 151, 153, 157, 158, 159, 160–161, 162, 169, 170, 172, 178, 179, 180, 181, 182, 183, 189, *191*, 192, *193*–194, 196, 197, 200
Hendrix, Jimi, 13
Henrik Johansson, 17, 99, 157, 169, 170, 180, 192
Heroes (TV series), 164n.1
Herter, Cosima, 196. *See also* Cosima Niehaus
Hewett, Richard, 52
Hitler, Adolf, 105
Hoffman, E.T.A., 64
Hollywood labour and crafts fields, 55–56n.5
Huang, S.L., 88, 89, 91n.14
Human Genome Project, 115–116
Humann, Heather Duerre, 141–142
Huxley, Aldous, 17, 100, 101, 102, 103–104, 105, 106, 108, 109n.2
Huxley, Thomas Henry, 100, 102

I
Ibsen, Henrik, 53
Industrial Light & Magic, 61
Ira (clone), 105, 141, 190
Irigaray, Luce, 169, 181
Irwin, John T., 164, 165n.5
The Island of Doctor Moreau (novel), 17, 99–100, 101–103, 108, 136, 149, 155, 163, 176

J
Jentsch, Ernst, 66
Jesse, 141, 161
La Jetée, 164n.1
Jurassic Park (1993 film), 15–16, 61, *62*, 63, 64, 66, 68, 69–73, 74; Mr DNA video in 69–73

K
Kapur, Isabella, 64
Katja Obinger, 3, 43, 44, 62, 81, 119–120, 138, 151, *158*, 181
Keller, Evelyn Fox, 117
Kendall Malone, 98, 105, 140, 163, 178, 189, 201n.1

Kendra Dupree, 13, 189
Kira Manning, 4, 5, 35, 36, 75, 98, 101, 103, 104, 108, 132, 136, 137, 142, 153, 155, *156*, 157, 160, 161, 170, 171, 178, 181, 187, 189, 190, 194
Kirby, Michael, 42, 43
Kittler, Friedrich, 68, 69, 70, 71
Kleist, Heinrich von, 36
Krystal Goderitch, 6, 43, 81, 97, 120, 151, 195
Kuleshov, Lev, 49–50

L
LaValley, Albert, 60, 67, 68
Lacan, Jacques, 117, 125n.3, 165n.6
Leda clones, 17, 48, 75, 97, 98, 99, 101, 103, 104, 105, 107, 108, 132, 133, 134, 138–139, 140, 141, 142, 144, 154, 163, 170, 178, 190. *See also* Project Leda
Lepore, Jill, 59, 67
Levin, Ira, 17, 100, 105, 106, 108
Levine, Alex, 144n.2, 197
Lifespring, 21n.4
Loofbourow, Lili, 20, 20n.3, 44, 45, 52, 156
Los Angeles Times, 42
Lost (TV series), 59, 60
Lost Girl, 108
Lumière brothers, 37n.7
Lynch, Stephen, 84, 85, 196–197

M
Malabou, Catherine, 20, 140, 172–173, 177–178, 182
Manning, Erin, 35, 54, 55
Manovich, Lev, 68–69
Manson, Graeme, 28, 29, 30, 31, 32, 51, 98, 196, 197
Mark Rollins, 4, 136, 137, 194
Markham, Annette, 91n.3
Maslany, Tatiana, 3, 4, 9, 14, 15, 16, 18, 20, 28, 29–31, *32–33*, 35, 36, 42, 43, 44, 45, 46, 51, 52, 53, 54, 55, 59, 60, *63*, 74, *75*, 76, 79, 81, 82, 84, 85, 86, 89, 90n.1, 97, 108, 113, 120, 138, 140, 141, 142, 151, 153, 194, 195, 197–198, 200; and Emmy Award win 3, 42, 79
Mathijs, Ernest, 107, 108, 109n.3
May, Emmeline, 87–88
Mbembe, Achille, 148, 162
McConnell, Frank, D., 102–103
McGowan, Todd, 154
McLeod, Randall, 54
McLuhan, Marshall, 73
McMillan, Graeme, 120
Méliès, Georges, 14, 27, 32, 33, 37
Mendel, Gregor, 115
Mengele, Josef, 105
Meriwether, Elizabeth, 53
Method acting, 8, 41–42, 44, 47, 55n.3
Metz, Christian, 68, 69
Mills, Hayley, 27, 28
M.K. *See* Veera 'M.K.' Suominen
montage and editing in filmmaking, 31, 32, 37n.4–5, 42, 45, 49–50, 69
motion-captured performance, 48–49
Mrs S, 4, 5, 19, 35, 98, 99, 104, 105, 133, 136, 161, 170, 171, 181. *See also* Siobhan Sadler
Murders in the Rue Morgue (novel), 164n.3
Myriad Genetics, 125n.3

N
Nancy, Jean-Luc, 130, 179, 180
Naremore, James, 45
Neolution(ists), 7, 8, 13, 18, 65, 97, 101, 102, 103, 108, 118, 123, 129, 130, 132, 133, 134, 135, 136, 137, 139, 140, 142, 143, 182, 187, 189
Neroni, Hilary, 5, 17, 189
New Atlantis, 8
The New Organon, 8
New York Times, 120
New York Times Magazine, 20, 44
New Yorker, The, 59, 67, 91n.8
Ngai, Sianne, 73
Nietzsche, Friedrich, 122

North, Daniel, 6
Nussbaum, Emily, 142

O

Odyssey, 5, 12
Olivier Duval, 11–12
The Origin of Species, 3, 7, 102, 115
Orphan Black, 3, 4, 5, 6, 7, 11, 15, 17–18, 29, 34, 36, 37, 42, 43, 45, 47, 48, 50, 51, 52, 53, 54, 55, 59, 60, 61, 62, 63, 64, 67, 68, 69, *74*, *75*, 79–88, 97, 99, 101, 104, 105, 106, 118, 124, 130, 133, 136, 139, 140, 143, 148, 149, 151, 152, 154, *156*, *158*, *159–160*, 162, 172, 178, 187; animation in 15–16; biopolitical content of 7, 10–11, 12, 18–19, 115, 116, 119, 123–124, 125n.3, 130–132, 140, 147–148, 152, 153, 162–163, 170, 172, 173, 177, 182–183, 190, 195; Canadian origins of 12, 103, 196; camera technique of 15, 29, 51–53, 54, 55, 83, 84, 85, 142, 172, 193–194, 197; 'Clone Club' 6, 8, 12, 18, 20, 79, 107, 118, 124, 130, 137, 138, 144, 149, 173, 177, 179, 180, 181, 183, 187, 190; and cloning 53, 55, 61, 62, 63, 97, 101, 113, 119, 120, 136, 148, 155; comics 174; critical reaction to 59; dialogue in 134, 138, 143, 152, 154; and cultural influences 174, 177; and DNA 4, 13, 17, 18, 19, 74, 76, 113–118, 119–120, 122, 123–124, 148, 149, 151, 152, 153, 159, 163, 189–190; and filiation 17, 98–99, 171, 179, 181–182, 188, 190, 200–201n.1; gender symbolism 5, 6, 7, 9–11, 12, 13, 14, 16, 18, 20, 43, 49, 82, 149, 183; make-up 3, 7, 16, 17, 19, 31, 33, 43, 47, 49, 50, 52, 55, 56, 60, 61, 64, 67, 80, 81, 82, 84–85, 86, 88, 91, 97, 98, 195–196, 197; genesis of series 196; and neuroscience 14, 81, 86, 90; parodies of 6; performances in 3, 10, 15, 18, 30, 37, 41–55, 84, 85, 141, 197–198, 200; production team 16–17, 19, 51, 83–84; and prosopagnosia 16, 79–81, 84, 86–90, 90n.2; and race 12–13; and religion 161, 188, 192, 201n.4; *sestras* 6, 12, 13, 15, 18, 97, 104, 106, 107, 108, 129, 130, 133–134, 137, 138–139, 141, 142, 143–144, 160, 161, 162, 163, 179, 182, 190; and sexuality 12, 20n.2, 43, 80, 82, 83, 89, 151, 169, 170, 171, 190, 192; special effects 4, 14–16, 28–29, 32–*33*, 45, 62, 63, 66–67, 75, *76*, 83, 84, 142, *199*; and technology 12, 13–14, 15, 49, 65; web responses to 79, 87–89, 90–91n.3, 124n.1, 144; and women's health and reproductive issues 7, 11, 12, 13, 14, 64, 147, 153, 155, 157, 160, 161, 163, 169, 171, 172, 178, 181, 182, 191, 192, 194, 200
Orphan Black characters. *See* Alison Hendrix, Aynsley Norris, Beth Childs, Bonnie Johansson, Cosima Niehaus, Delphine Cormier, Donnie Hendrix, Dr Aldous Leekie, Dr Nealon, Dr Virginia Coady, Ethan Duncan, Evie Cho, Felix, Ferdinand, Gracie Johanssen, Helena, Henrik Johansson, Ira, Jesse, Katja Obinger, Kendall Malone, Kendra Dupree, Kira Manning, Krystal Goderitch, Mark Rollins, Mrs S/Siobhan Sadler, Olivier Duval, Paul, P.T. Westmorland, Rachel Duncan, Sarah Manning, Susan Duncan, Tomas, Tony, Veera 'M.K.' Suominen, Vic Schmidt, Yanis
Orphan Black clones, 4, 5, 6–7, 8, 10, 11, 12, 13, 14, 15, 16, 17, 18, 19, 20n.1, 28, 29, 30, 31, 32, *33*, 34, 35, 36, 37 *46*, 48, 51, 53, 54, 55, 59, 61, 62, *63*, 64, 69, 75, 76, 79, 80, 81, 82–83, 84, 85, 86, 88–89, 90, 106, 107, 113, 114, 118, 119–*121*, 122, 124, 129, 130, 134–138, 140–144,

151–155, 157, 170–178, 181, 183, 188–189, 191, 194, 200
Outerbridge, Peter, 147, 197

P
The Parent Trap (1961), 15, 27, 29, 32, 35
Paradis, James, 100
The Passion of Joan of Arc (1928 film), 196, 197
Paul (character), 86, 134
Pearl, Sharrona, 4, 16
Pence, Gregory, 12, 148
Poe, Edgar Allan, 149, 152, 155, 164n.3–4
Porzak, Simon, 4, 15, 16, 187
Proceedings of the German Botanical Society, 115
Project Castor, 7, 97, 108. *See also* Castor clones
Project Leda, 129, 141, 154, 188. *See also* Leda clones
Proletheans, 7, 83, 97, 106, 108, 137, 140, 149, 153, 157, 160, 170, 171, 179, 181, 194, 197
prosopagnosia, 16, 79–81, 84, 86–90, 90n.2, 91n.8
P.T. Westmorland, 18, 97, 99, 100, 102, 103, 108, 129, 133, 134, 136, 137, 138, 139, 140, 141, 142, 149, 157, 170, 188, 189, 199, 200
"Protest" (poem), 8

Q
Québec sovereignty movement, 144n.1

R
Rachel Duncan, 6, 12, 18, 19, 20, 43, 60, 81, 85, 97, 98, 99, 101, 102, 108, 119, 122, 130, 133, 134, *135*, 136, 137, 138, 139, 142, 143, 151, 155, 156, 157, 160, 161, 163, 169, 170, 171, *172*, 174, 177, 180, 182, 183, 187, 188, 189, 191, *192*, 193, 194, *199*, 200, 200n.1, 201n.3
Randolph, Lynn, *182*
Rank, Otto, 64, 108n.1
Reddit, 81, 88, 91n.3
Redfield, Marc, 129, 133
Reeves, Matt, 60
ReGenesis, 12, 147
La Règle du jeu (1939), 31–32, 36, 37n.6
The Reject: Community, Politics, and Religion after the Subject, 173, 179
Renoir, Jean, 14, 31–32, 36
La Reproduction Interdite (novel), 19, 172, 173–176, 179, 181. *See also* French sci-fi cloning fiction
Ringer, 3
Roach, Joseph, 46–47
Robins, Elizabeth, 53–54
Rosenheim, Shawn, 164n.3–4, 165n.5
The Rules of the Game (1939). *See Règle du jeu, La*
R.U.R. (Rossum's Universal Robots), 48
Rushing, Robert A., 8, 11, 17, 19, 37n.1, 82

S
Sacks, Oliver, 91n.8
Sarah Manning, 3, 4, 5, 6, 7, 8, 9, 10, 13, 18, 19, 20n.3, 29, 30, *33*, 35, 43, 44, 54, 55, 60, 62, 64, 74, *75*, 76, 81, 82, 83, 84, 86, 89, 97, 98, 99, 101, 103, 104, 105, 106, 108, 113, 114, 118, 119, *121*, 122, 123, 124, 129, 130, 132, 133, 136, 137, 138, 139, 140, 141, 142–143, 144, 144n.2, 151, 153, 154, 155, 156, 157, 160, 161, 163, 169, 170, 171, 172, 178, 180–181, 182, 183, 187, 189, 190, 191, 192, 193, 194, 197, 198, 200
Schmitt, Carl, 11
science fiction and technology, 66–67
Scott, Geoff, 45, 142
The Secret Circle, 108
Serkis, Andy, 49, 56n.5
Sex and the City (TV series), 6, 20n.3
Shacklock, Zoë, 45, 142
Sherlock Holmes, 19, *149*–150, 151, 155, 157, 162, 164
Silvio, Teri, 74
Simians, Cyborgs, and Women: The Reinvention of Nature, 8, 182

Simulacra and Simulation, 68
Singh, Simon, 165n.10
Siobhan Sadler, 4, 5, 98, 170, 189–190, 191. *See also* Mrs S
Slate, Jenny, 195
Sloterdijk, Peter, 130
Sokolowski, Sandy, 84, 85
special effects, 68–69
Speculum of the Other Woman, 169
Spielberg, Steven, 15, 71
Stanislavsky, Konstantin, 15, 47
Stewart, Garrett, 74
Stout, John, C., 8, 17
Strachey, James, 99
Strasberg, Lee, 41, 44
The Student of Prague, 64
Stutsman, Staci, 43
Susan Duncan, 17, 98, 99, 103, 133, 134, 135, 136, 137, 142, 170, 182, 188, 189, 200n.1

T
Tait, T. Colin, 43
Tanner, Jessica, 8, 17, 18
TechnoDolly, 15, 29, 51–53, 54, 55, 83, 84, 85, 120, 142
technology: film and digital, 69, 73, 74
Terminator film series, 164n.1
Théorie de la démarche (Theory of the Gait), 37n.8
Thompson, Kristin, 37n.4
Todorov, Tzvetan, 66
Tomas, 153, 157, 174, 179, 194, 201n.1
Tony (clone), 6, 12, 81, 119, 122, 151, 159, 171, 190
Topside, 7, 8, 133, 170, 183, 187
Travelers (TV series), 12, 164n.1
Trump, Donald, 153, 162
Truong, Jean-Michel, 19, 172, 173, 174, 175, 177, 179, 181
20,000 Leagues Under the Sea (novel), 66
Twitter, 43, 144

U
Unleashed, 107

V
Väliaho, Pasi, 147, 162
Veera 'M.K.' Suominen, 97 120, 143, 180, 182, 189, 201n.4
Verne, Jules, 66
Vic Schmidt, 10, 13
Vigenère cipher, 154, 155
Violence (book), 180
The Vital Illusion, 67

W
Walton, Karen, 109n.3
Washington Post, 42
Watson, James, 115, 117
Wells, H.G., 17, 99–100, 101, 102, 103, 108
Westworld (TV series), 48
What We Talk about When We Talk about Clone Club, 12
Where Late the Sweet Birds Sang, 17, 100, 105, 106
Wilcox, Ella Wheeler, 8
Wilhelm, Kate, 17, 100, 105, 106, 108
Wolf, Mark J.P., 49
Wolfinger, Emily, 91n.3
World War II, 165n.10

X
The X-Files (TV series), 99

Y
Yanis, 103, 108
YouTube, 29, 32, 50, 195

Z
Žižek, Slavoj, 162, 180

www.ingramcontent.com/pod-product-compliance
Lightning Source LLC
LaVergne TN
LVHW080831240826
846232LV00015B/56